Spin Waves

Daniel D. Stancil • Anil Prabhakar

Spin Waves

Problems and Solutions

 Springer

Daniel D. Stancil
Department of Electrical
and Computer Engineering
North Carolina State University
Raleigh, NC, USA

Anil Prabhakar
Department of Electrical Engineering
IIT Madras
Chennai, Tamil Nadu, India

ISBN 978-3-030-68584-3   ISBN 978-3-030-68582-9   (eBook)
https://doi.org/10.1007/978-3-030-68582-9

This Springer imprint is published by the registered company Springer Nature Switzerland AG
The registered company address is: Gewerbestrasse 11, 6330 Cham, Switzerland

To our students

Preface

Research on spin waves over the past decade has been facilitated by improvements in materials, and in nano-fabrication methods. To support the pedagogical interest of the academic and research community, we began compiling a solutions manual for the chapter problems in *Spin Waves: Theory and Applications*. However, as the manuscript developed, we felt that it would be more useful if we expanded the content to include a short summary of key concepts and ideas needed to solve the problems. As a result, the present work can be used as a stand-alone volume, though we do recommend that the reader treat this volume as a complementary text to *Spin Waves*.

The references at the end of each chapter provide contextual, and sometimes historical, information on spin waves. To enhance the readability of the problems without having to continually refer back to the corresponding portions of *Spin Waves*, we have adopted a different style of writing, with explanatory text interspersed with the problem statement and solutions. The problems have also been substantially re-worded, and we used this opportunity to clarify the statement of some problems, and add a few new problems. We expect that most of our readers will be familiar with programming languages, and we have included Python code to assist them in generating some of the plots that accompany the solutions to the problems.

The structure of each chapter consists of short sections introducing and summarizing key concepts interspersed with problems and solutions pertaining to the material. To facilitate distinguishing between the solutions themselves and the problem statements and summary sections, vectors are denoted by boldface type in the summaries and problem statements, e.g., $\mathbf{H}$, but italic type with an overbar is used in the solutions, as $\bar{H}$. Similarly, tensors are denoted by boldface type with an overbar in the summaries and problem statements, e.g., $\bar{\boldsymbol{\mu}}$, but italic type with a double overbar in the solutions, as $\bar{\bar{\mu}}$. References to pertinent sections of text in *Spin Waves* are provided, encouraging readers to explore concepts in further depth.

As with the parent *Spin Waves* text, this book can be divided into three major parts. The first is comprised of Chaps. 1–3 and is concerned with the physics of magnetism in magnetic insulators. The principal goals of these chapters are to provide a basic understanding of the microscopic origins of magnetism and exchange-dominated spin waves, motivate the equation of motion for the macroscopic magnetization, and to construct appropriate susceptibility models to describe the linear responses of magnetic materials to magnetic fields. The second part, Chaps. 5–8, focuses on magnetostatic modes and dipolar spin waves, their properties, how to excite them, and how they interact with light. Chapter 4 serves as a bridge between these two parts by discussing how the susceptibility models from Chap. 3 can be used with Maxwell's equations to describe electromagnetic and magnetoquasistatic waves in dispersive anisotropic media. Finally, Chaps. 9 and 10 treat nonlinear phenomena and advanced applications of spin wave excitations.

We hope you find the style adopted in this book to be useful, and it encourages you to learn through problem-solving, even as you pursue a comprehensive exploration of *Spin Waves*.

Raleigh, USA   Daniel D. Stancil
Chennai, India   Anil Prabhakar

Acknowledgments

We are indebted to a number of people for helpful discussions and comments on portions of this book.

The accuracy and readability of the initial work, *Theory of Magnetostatic Waves*, were improved considerably by comments and suggestions from N. Bilaniuk, N. E. Buris, S. H. Charap, D. J. Halchin, J. F. Kauffman, T. D. Poston, A. Renema, S. D. Silliman, M. B. Steer, and F. J. Tischer. In addition, the previous volume, *Spin Waves: Theory and Applications*, benefited from our interactions with C. E. Patton, P. E. Wigen, and A. N. Slavin on nonlinear excitations, auto-oscillations, and soliton formation; from discussions with M. Widom on quantum mechanics; and from comments and suggestions relating to spin-transfer torques from J. C. Slonczewski. Of course, the remaining errors and idiosyncrasies are ours.

One of us (DDS) would particularly like to acknowledge his late mentor, colleague, and friend, Prof. F. R. Morgenthaler, for teaching him much of the material in this book. He is also grateful to Kathy for her love, support, and patience. AP remains grateful to DDS for introducing him to the rich physics of magnetostatics, spin waves, and magneto-optical interactions. AP also thanks his wife, Namita, for her encouragement and her indulgence during the many stages of this manuscript, and his former students M. Malathi, N. Kumar, and G. Venkat, who have been co-voyagers on his explorations in spin waves over the past decade.

The solutions to the problems were collated with assistance from our former students, N. E. Buris, C. Shivpriya, and R. Malpani. AP also acknowledges the help he received in typesetting the solutions from many of his student teaching assistants and staff.

Finally, it has been a pleasure to work with Zachary Evenson, Kavitha Palanisamy, Adelheid Duhm and their capable team at Springer.

Contents

Introduction to Magnetism

We begin our study of spin waves with a discussion of the origins of magnetism. We start by building some intuition about classical angular momentum with the motion of a spinning top, then move on to a quantum mechanical treatment of angular momentum. We end the chapter by discussing the use of Hund's Rules to find the ground states of ions important to magnetism.

The dynamics of a spinning top using classical mechanics is considered in Problem 1.1, while Problems 1.2–1.5 explore angular momentum in quantum mechanics. Problems 1.6 and 1.7 concern the ground states of ions important to magnetism, and Problem 1.8 uses the Hund Rule ground state of Fe^{3+} to calculate the maximum magnetic moment per unit volume at $0\,K$ of the technologically important material Yttrium Iron Garnet.

1.1 Analogy of the Spinning Top

The magnetic properties of materials are the result of orbital and spin angular momentum of electrons. Although on the atomic scale angular momentum must be treated quantum mechanically, the macroscopic dynamics of the magnetization can be understood with the help of classical models.[1]

In particular, the classical equation of motion for the angular momentum $\mathbf{J}$ of a charged particle can be written as

$$\frac{d\mathbf{J}}{dt} = \mathbf{\Omega} \times \mathbf{J}, \tag{1.1}$$

---

[1] See, for example, *The Feynman Lectures* Vol. I [1] and Goldstein [2].

© The Author(s), under exclusive license to Springer Nature Switzerland AG 2021
D. D. Stancil and A. Prabhakar, *Spin Waves*,
https://doi.org/10.1007/978-3-030-68582-9_1

where

$$\mathbf{\Omega} = -\gamma \mathbf{B}, \tag{1.2}$$

$\mathbf{B}$ is an applied magnetic flux density, and γ is the *gyromagnetic ratio*, a constant that relates the electron angular momentum to the resulting magnetic moment μ:

$$\mu = \gamma \mathbf{J}. \tag{1.3}$$

In terms of fundamental parameters, the gyromagnetic ratio is given by

$$\gamma = g \frac{q}{2m_q}, \tag{1.4}$$

where m_q is the mass of the particle with charge q (typically an electron), and g is a factor that depends on the type of angular momentum, and is called the *Landé g factor*. Classically it has the value 1 and represents orbital angular momentum. The equations are also consistent with a quantum mechanical treatment if g is redefined. In particular, when the angular momentum is treated quantum mechanically, g has the value 1 for orbital angular momentum, 2 for spin angular momentum, and other values for mixtures of spin and orbital angular momentum. For mixtures, the Landé g factor is given by

$$g = \frac{3}{2} + \frac{S(S+1) - L(L+1)}{2J(J+1)}, \tag{1.5}$$

where S is the total spin, L is the total orbital angular momentum, and J is the total angular momentum. Intuition about the behavior of solutions to this equation of motion can be gained by noting that the equation of motion for a top spinning in a gravitational field can also be written in the form of (1.1) but with

$$\mathbf{\Omega} = \omega_P \mathrm{sgn}(\mathbf{d} \cdot \mathbf{J}) \hat{z}, \tag{1.6}$$

where $\mathrm{sgn}(x)$ gives the algebraic sign of x and $\mathbf{d}$ is the vector from the origin to the center of mass of the top. The coefficient is given by

$$\omega_P = mg_0 \, d/J, \tag{1.7}$$

where m is the total mass of the top and g_0 is the gravitational acceleration (9.8 m/s^2). The angular momentum can be related to the rate at which the top is spinning about its axis by

$$|\mathbf{J}| = I\omega_0, \tag{1.8}$$

where ω_0 is the angular velocity of the spinning top, and I is the mass moment of inertia defined by

$$I = \int_V r^2 \rho \mathrm{d}v. \tag{1.9}$$

Here r is the distance between the axis of rotation and the mass element $dm = \rho dv$, and ρ is the mass density of the top.

Because the equations of motion are the same, we can gain some insight into the dynamics of the charged particle from what we know of the behavior of a top. In the case of a top, the mass plays the role of the charge, and the gravitational field plays a similar role to the magnetic flux density. When a top spins, it precesses about the vertical direction, and as it spins down it precesses faster and faster until it falls over.

The frequency of the precession of a top is $\omega_P = mg_0d/J$, while that for a charged particle is $\omega_P = |\gamma|B$. Although we can't readily vary the strength of the gravitational field, we see that both frequencies increase with an increasing field (gravitational field in one case, and magnetic field in the other). Further, the inverse dependence on J describes the increase with precession frequency as the top spins down. This leads to an important observation of a key difference between the models: in the case of the charged particle in a magnetic field, the precession frequency depends on the *type* of angular momentum (orbital, spin, or a mixture), but not on the *magnitude* of the angular momentum. Although there are some differences, the mental picture of a precessing top gives an intuition for the precession of a charged particle with spin.

Problem 1.1 Consider a top in the shape of a right circular cone with height h and maximum radius R.

(a) Show that the mass moment of inertia about the symmetry axis of the top is given by

$$I = \frac{3mR^2}{10},\qquad (1.10)$$

and that the center of mass is located on the axis of the cone a distance $d = 3h/4$ from the apex.

(b) Using the results of part (a), show that the angular precession frequency is given by

$$\omega_P = \frac{5hg_0}{2R^2\omega_0}.\qquad (1.11)$$

(c) If $h = 0.016\,\mathrm{m}$, $R = 0.020\,\mathrm{m}$, and the top is spinning at 1500 RPM, calculate $f_P = \omega_P/2\pi$.

(d) How will the precession frequency change as the top slows down? What will be the effect of spinning the top in the opposite direction?

Solution 1.1

(a) For the top shown in Fig. 1.1, $I = \int r^2 dm$ where $dm = \rho dv = \frac{m}{V} r \, dr \, d\phi, dz$

$$\therefore I = \frac{m}{V} \int_0^h dz \int_0^{r_{max}} r^3 dr \int_0^{2\pi} d\phi = \frac{2\pi m}{V} \int_0^h dz \left.\frac{r^4}{4}\right|_0^{r_{max}}\qquad (1.12)$$

Fig. 1.1 Geometry of a spinning top

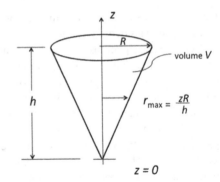

where $r_{max} = \frac{zR}{h}$. Hence,

$$I = \frac{\pi m R^4}{2V h^4} \int_0^h z^4 dz = \frac{\pi m R^4 h}{10V}. \tag{1.13}$$

The volume of the cone is

$$V = \int_0^h dz \int_0^{r_{max}} r\, dr \int_0^{2\pi} d\phi = 2\pi \int_0^h dz \left. \frac{r^2}{2} \right|_0^{r_{max}} \tag{1.14}$$

$$= \frac{\pi R^2}{h^2} \int_0^h z^2 dz = \frac{\pi R^2 h}{3}. \tag{1.15}$$

Substituting (1.15) into (1.13) yields $\boxed{I = \dfrac{3m R^2}{10}}$.

The center of mass is calculated as

$$d_{CM} = \frac{1}{m} \int z\, dm,$$

where $dm = \rho \pi r^2 dz$ is the differential mass of a disc of radius r and thickness dz at a height z from the tip of the cone, as shown in Fig. 1.2. From simple geometrical considerations, we have $r = \frac{zR}{h}$ and $\rho = m/V = \frac{3m}{\pi R^2 h}$.

$$\therefore \quad d_{CM} = \frac{1}{m} \int_0^h z \frac{3m}{\pi R^2 h} \pi \left(\frac{zR}{h} \right)^2 dz \quad \Rightarrow \quad \boxed{d_{CM} = \frac{3h}{4}}.$$

(b)

$$\omega_P = \frac{m g_0 d_{CM}}{I \omega_0} = \frac{m g_0}{\omega_0} \frac{3h}{4} \frac{10}{3m R^2}$$

$$\boxed{\omega_P = \frac{5 h g_0}{2 R^2 \omega_0}.}$$

Fig. 1.2 Center of mass for a spinning top

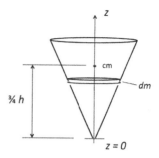

(c) We have $h = 0.016$ m, $R = 0.02$ m, and speed = 1500 RPM. Thus

$$\omega_0 = 1500\frac{\text{rev}}{\text{min}} \cdot \frac{\text{min}}{60\,\text{s}} \cdot \frac{2\pi\,\text{rad}}{\text{rev}} = 157.1\ \text{rad/s}.$$

$$f_P = \frac{\omega_P}{2\pi} = \frac{5hg_0}{4\pi R^2\omega_0} = \frac{5(0.016)9.8}{4\pi(0.02)^2157.1}$$

$$\boxed{f_P \approx 1\ \text{rev/s}}.$$

(d) As the top slows down, f_P *increases*. Reversing the direction of the spin will reverse the direction of precession.

1.2 Angular Momentum in Quantum Mechanics

The model for spin motion that we have described is based on classical physics, and is a good description of the average behavior of large numbers of spins. To get a better understanding of the dynamics of an individual spin, however, we must treat angular momentum quantum mechanically.[2]

The transition from classical to quantum mechanics is made by replacing classical quantities with *operators* that are then applied to the *wave function* of the system. The expected values of the operators then represent the allowed values of the quantities in the quantum mechanical system.

In the case of angular momentum, the operators have the interesting property that they do not commute, i.e.,

$$(L_xL_y - L_yL_x)\psi \equiv [L_x, L_y]\psi \neq 0. \tag{1.16}$$

[2]Detailed treatments of angular momentum in quantum mechanics can be found, for example, in Eisberg [3], *The Feynman Lectures* Vol. III [4], Merzbacher [5], and Schiff [6].

The square bracket above is referred to as a *commutator*. The value of a commutator is an important property providing information about the relationship between the operators in the commutator. It turns out that for operators that commute (i.e., the value of the commutator is zero), wave functions can be found that are eigenvectors of both operators. Said differently, if A and B are operators on the wave function ψ, then we can find a wave function that satisfies eigenvalue equations for both operators; i.e., $A\psi = a\psi$ and $B\psi = b\psi$ can both be satisfied by the same wave function, for specific values of a and b. The values of a, b are referred to as "good quantum numbers" for the system.

For an operator to properly represent the angular momentum vector, the components of the operator must satisfy the follow commutation relations:

$$[L_x, L_y] = i\hbar L_z, \quad [L_y, L_z] = i\hbar L_x, \quad \text{and} \quad [L_z, L_x] = i\hbar L_y. \tag{1.17}$$

These relationships can be compactly represented by the vector cross-product

$$\mathbf{L} \times \mathbf{L} = i\hbar \mathbf{L}. \tag{1.18}$$

An operator must satisfy this equation to be a proper *angular momentum operator*. Classically, the vector cross-product of a vector with itself is equal to zero. The cross-product does not vanish in this case because the components of the vector do not commute.

It turns out that since the components of the angular momentum vector do not commute, only one can be associated with a good quantum number. By convention, we usually choose this to be the z component. The associated eigenvalue equation is

$$L_z \psi = m\hbar \psi \tag{1.19}$$

where m is an integer. Although L_z does not commute with the other component operators, it turns out that it does commute with $L^2 = \mathbf{L} \cdot \mathbf{L} = L_x^2 + L_y^2 + L_z^2$. The associated eigenvalue equation for L^2 is usually written

$$L^2 \psi = \hbar^2 l(l+1)\psi, \tag{1.20}$$

where m and l are related by

$$-l \leq m \leq l, \tag{1.21}$$

and m and l are integers for orbital angular momentum. Similar relations hold for spin angular momentum $\mathbf{S}$ except that the total angular momentum quantum number (referred to as s instead of l) can be a half-integer as well as an integer.

It is also useful to define the angular momentum raising and lowering operators:

$$L_+ = L_x + iL_y, \quad L_- = L_x - iL_y. \tag{1.22}$$

To see where the names come from, it can be shown that if ψ_m is an eigenfunction of L_z with eigenvalue $\hbar m$, then $(L_+ \psi_m)$ is an eigenfunction of L_z with eigenvalue

$\hbar(m + 1)$. The effect of L_+ is therefore to raise the z component of angular momentum. Similarly, $(L_-\psi_m)$ is an eigenfunction of L_z with eigenvalue $\hbar(m - 1)$, and therefore lowers the z component of angular momentum.

Problem 1.2 Using $[L_x, L_y] = i\hbar L_z$, $[L_y, L_z] = i\hbar L_x$ and $[L_z, L_x] = i\hbar L_y$ verify

$$[L_z, L_+] = \hbar L_+ \tag{1.23a}$$
$$[L_z, L_-] = -\hbar L_- \tag{1.23b}$$
$$[L_+, L_-] = 2\hbar L_z \tag{1.23c}$$

and

$$[L^2, L_x] = [L^2, L_y] = [L^2, L_z] = [L^2, L_+] = [L^2, L_-] = 0. \tag{1.24}$$

Solution 1.2

$$
\begin{aligned}
[L_z, L_+] &= L_z (L_x + iL_y) - (L_x + iL_y) L_z \\
&= L_z L_x - L_x L_z + iL_z L_y - iL_y L_z \\
&= [L_z, L_x] + i [L_z, L_y] \\
&= i\hbar L_y + i (-i\hbar L_x) \\
&= \hbar (L_x + iL_y) \\
&= \hbar L_+
\end{aligned}
$$

Hence,

$$\boxed{[L_z, L_+] = \hbar L_+}$$

$$
\begin{aligned}
[L_z, L_-] &= L_z (L_x - iL_y) - (L_x - iL_y) L_z \\
&= L_z L_x - L_x L_z - iL_z L_y + iL_y L_z \\
&= [L_z, L_x] + i [L_y, L_z] \\
&= i\hbar L_y + i (i\hbar L_x) \\
&= \hbar (-L_x + iL_y) \\
&= -\hbar L_-
\end{aligned}
$$

Hence,

$$\boxed{\left[L_z, L_-\right] = -\hbar L_-}$$

$$
\begin{aligned}
\left[L_+, L_-\right] &= \left(L_x + iL_y\right)\left(L_x - iL_y\right) - \left(L_x - iL_y\right)\left(L_x + iL_y\right) \\
&= L_x^2 + iL_yL_x - iL_xL_y + L_y^2 - \left(L_x^2 - iL_yL_x + iL_xL_y + L_y^2\right) \\
&= i\left[L_y, L_x\right] - i\left[L_x, L_y\right] \\
&= -2i\left[L_x, L_y\right] \\
&= -2i\left(i\hbar L_z\right) \\
&= 2\hbar L_z
\end{aligned}
$$

Hence,

$$\boxed{\left[L_+, L_-\right] = 2\hbar L_z}$$

$$
\begin{aligned}
\left[L^2, L_x\right] &= \left(L_x^2 + L_y^2 + L_z^2\right)L_x - L_x\left(L_x^2 + L_y^2 + L_z^2\right) \\
&= L_x^3 + L_y^2L_x + L_z^2L_x - L_x^3 - L_xL_y^2 - L_xL_z^2 \\
&= L_y^2L_x + L_z^2L_x - L_xL_yL_y - L_xL_zL_z
\end{aligned}
$$

But,

$$L_xL_y = L_yL_x + i\hbar L_z$$

and

$$L_xL_z = L_zL_x - i\hbar L_y$$

Thus

$$
\begin{aligned}
\left[L^2, L_x\right] &= L_y^2L_x + L_z^2L_x - \left(L_yL_x + i\hbar L_z\right)L_y - \left(L_zL_x - i\hbar L_y\right)L_z \\
&= L_y^2L_x + L_z^2L_x - L_yL_xL_y - i\hbar L_zL_y - L_zL_xL_z + i\hbar L_yL_z \\
&= L_y^2L_x + L_z^2L_x - L_y\left(L_yL_x + i\hbar L_z\right) - i\hbar L_zL_y - L_z\left(L_zL_x - i\hbar L_y\right) + i\hbar L_yL_z \\
&= L_y^2L_x + L_z^2L_x - L_y^2L_x - i\hbar L_yL_z - i\hbar L_zL_y - L_z^2L_x + i\hbar L_zL_y + i\hbar L_yL_z \\
&= 0
\end{aligned}
$$

Following a similar procedure $[L^2, L_y] = [L^2, L_z] = 0$. Since, L_+ and L_- are linear combinations of L_x and L_y, it follows from the above results that

$$[L^2, L_+] = [L^2, L_-] = 0$$

Hence,

$$\boxed{[L^2, L_x] = [L^2, L_y] = [L^2, L_z] = [L^2, L_+] = [L^2, L_-] = 0}$$

Problem 1.3 If

$$L^2\psi = \hbar^2 l(l+1)\psi \tag{1.25}$$

and

$$L_-(L_-)^n\psi = 0, \tag{1.26}$$

show that $n = 2l$ and thus it follows that $-l \leq m \leq l$.

Solution 1.3

Assume ψ is a wave function with a maximum value of $m = l$. We then need to find the smallest value of n for which $L_- (L_-)^n \psi = 0$. The minimum value of m would then be given by

$$L_z (L_-)^n \psi = \hbar m_{min} (L_-)^n \psi$$

The following relation will be useful, and can be verified by substituting the definitions of $L_\pm$ and simplifying (see [7, Eq. (1.63)]):

$$L^2 = \frac{1}{2}(L_+L_- + L_-L_+) + L_z^2$$

But from Problem 1.2,
$$[L_+, L_-] = 2\hbar L_z$$
$$L_+L_- - L_-L_+ = 2\hbar L_z$$
$$\therefore L_-L_+ = L_+L_- - 2\hbar L_z$$
$$\Rightarrow L^2 = \frac{1}{2}(L_+L_- + L_+L_- - 2\hbar L_z) + L_z^2$$
$$= L_+L_- - \hbar L_z + L_z^2$$
$$= L_+L_- + L_z(L_z - \hbar).$$

Thus,

$$L^2 (L_-)^n \psi = [L_+L_- + L_z(L_z - \hbar)](L_-)^n \psi$$
$$= L_+L_- (L_-)^n \psi + L_z(L_z - \hbar)(L_-)^n \psi$$
$$= \hbar m_{min}(\hbar m_{min} - \hbar)(L_-)^n \psi$$

$$L^2 (L_-)^n \psi = \hbar^2 m_{min} (m_{min} - 1) (L_-)^n \psi. \qquad (1.27)$$

But we know

$$L^2 (L_-)^n \psi = \hbar^2 l (l + 1) (L_-)^n \psi. \qquad (1.28)$$

Since this must be true for all the valid values of m, comparison between (1.27) and (1.28) gives

$$m_{min} (m_{min} - 1) = l (l + 1)$$
$$m_{min}^2 - m_{min} = l^2 + l$$

This implies that $m_{min} = -l$

$$\boxed{\text{Thus, } n = 2l \text{ and } -l \leqslant m \leqslant l.}$$

Problem 1.4 Consider a two-particle wave function with total angular momentum $\mathbf{J} = \mathbf{J}_1 + \mathbf{J}_2$, where $\mathbf{J}_i$ is the angular momentum of the ith particle. Show that $[J^2, J_{1z}] \neq 0$ and $[J^2, J_{2z}] \neq 0$, so that eigenfunctions specified by the numbers m_1, m_2, j and m *cannot* be constructed.

Solution 1.4

$$[J^2, J_{1z}] = \left[(\bar{J}_1 + \bar{J}_2)^2, J_{1z} \right]$$
$$= [J_1^2 + J_2^2 + \bar{J}_1 \cdot \bar{J}_2 + \bar{J}_2 \cdot \bar{J}_1, J_{1z}]$$
$$= [J_1^2, J_{1z}] + [J_2^2, J_{1z}] + 2 [\bar{J}_1 \cdot \bar{J}_2, J_{1z}]$$

Here we have used $\bar{J}_1 \cdot \bar{J}_2 = \bar{J}_2 \cdot \bar{J}_1$ since $\bar{J}_1$ and $\bar{J}_2$ operate on different subspaces (i.e., different particles in this case). From the commutation relations in (1.24),

$$[J_1^2, J_{1z}] = 0.$$

Further, since $[\bar{J}_1, \bar{J}_2]$ operate on different subspaces

$$[J_2^2, J_{1z}] = 0.$$

Thus

$$[J^2, J_{1z}] = 2\bar{J}_2 \cdot [\bar{J}_1, J_{1z}]$$
$$= 2\bar{J}_2 \cdot [\hat{x} J_{1x} + \hat{y} J_{1y} + \hat{z} J_{1z}, J_{1z}]$$
$$= 2\bar{J}_2 \cdot (\hat{x} [J_{1x}, J_{1z}] + \hat{y} [J_{1y}, J_{1z}] + \hat{z} [J_{1z}, J_{1z}])$$

Using (1.17) this can be written as

$$
\begin{aligned}
\left[J^2, J_{1z}\right] &= 2\bar{J}_2 \cdot \left(\hat{x}\left(-i\hbar J_{1y}\right) + \hat{y} i \hbar J_{1x}\right) \\
&= 2i\hbar \bar{J}_2 \cdot \left(-\hat{x} J_{1y} + \hat{y} J_{1x}\right) \\
&\neq 0 \quad \text{in general}
\end{aligned}
$$

For a similar analysis it follows that $\left[J^2, J_{2z}\right] \neq 0$. As argued previously, a wave function can simultaneously be an eigenfunction of two operators only if the operators commute. Since the eigenvalues of J^2 and J_{1z} are $j\,(j+1)$ and m_1 respectively, $\left[J^2, J_{1z}\right] \neq 0$ means that j, m_1 cannot simultaneously be good quantum numbers. Similarly, $\left[J^2, J_{2z}\right] \neq 0$ means that j, m_2 cannot simultaneously be good quantum numbers. It follows that there cannot be an eigenfunction for which j, m_1, m_2 and $m = m_1 + m_2$ are simultaneously good quantum numbers.

Problem 1.5 Consider the spin eigenvalue equation

$$
S_z \chi_{s,m_s} = \hbar m_s \chi_{s,m_s}. \tag{1.29}
$$

For a single electron, m_s can only take on the two values $\pm\frac{1}{2}$. Since there are, in general, n eigenvalues and eigenvectors for an $n \times n$ matrix, S_z can be represented by a 2×2 matrix. In such a matrix representation, the total spin operator is defined by

$$
\mathbf{S} = \frac{\hbar}{2}\boldsymbol{\sigma}, \tag{1.30}
$$

where

$$
\boldsymbol{\sigma} = \hat{x}\sigma_x + \hat{y}\sigma_y + \hat{z}\sigma_z, \tag{1.31}
$$

and σ_x, σ_y and σ_z are the *Pauli spin matrices*:

$$
\sigma_x = \begin{bmatrix} 0 & 1 \\ 1 & 0 \end{bmatrix}, \quad
\sigma_y = \begin{bmatrix} 0 & -i \\ i & 0 \end{bmatrix}, \quad
\sigma_z = \begin{bmatrix} 1 & 0 \\ 0 & -1 \end{bmatrix}. \tag{1.32}
$$

(a) Verify that $\mathbf{S}$ is a true angular momentum operator by showing that

$$
\mathbf{S} \times \mathbf{S} = i\hbar \mathbf{S}. \tag{1.33}
$$

(b) Let

$$
\chi_\uparrow = \begin{bmatrix} 1 \\ 0 \end{bmatrix}, \quad
\chi_\downarrow = \begin{bmatrix} 0 \\ 1 \end{bmatrix}. \tag{1.34}
$$

Show that $\chi_\uparrow$ and $\chi_\downarrow$ satisfy

$$S_z \chi_\uparrow = \frac{\hbar}{2} \chi_\uparrow , \tag{1.35a}$$

$$S_z \chi_\downarrow = -\frac{\hbar}{2} \chi_\downarrow , \tag{1.35b}$$

and thus represent "spin-up" and "spin-down" states, respectively. These vectors are called *spinors*.

(c) Since spin coordinates are discrete, evaluating the expectation values of spin operators involves summations rather than integrals. Evaluate the following expectation values:

$$\langle S^2 \rangle = \chi_\downarrow^T S^2 \chi_\downarrow , \tag{1.36a}$$

$$\langle S_z \rangle = \chi_\downarrow^T S_z \chi_\downarrow , \tag{1.36b}$$

and

$$\langle S^2 \rangle = \chi_\uparrow^T S^2 \chi_\uparrow , \tag{1.36c}$$

$$\langle S_z \rangle = \chi_\uparrow^T S_z \chi_\uparrow , \tag{1.36d}$$

where the superscript T indicates the matrix transpose.

Solution 1.5

(a)

$$\bar{S} \times \bar{S} \stackrel{?}{=} i\hbar\bar{S}$$

$$\frac{\hbar}{2}\bar{\sigma} \times \frac{\hbar}{2}\bar{\sigma} \stackrel{?}{=} i\frac{\hbar^2}{2}\bar{\sigma}$$

$$\text{or,} \quad \bar{\sigma} \times \bar{\sigma} \stackrel{?}{=} 2i\bar{\sigma}$$

$$(\bar{\sigma} \times \bar{\sigma})_x \stackrel{?}{=} 2i\sigma_x$$

$$= \sigma_y\sigma_z - \sigma_z\sigma_y$$

$$= \begin{bmatrix} 0 & -i \\ i & 0 \end{bmatrix}\begin{bmatrix} 1 & 0 \\ 0 & -1 \end{bmatrix} - \begin{bmatrix} 1 & 0 \\ 0 & -1 \end{bmatrix}\begin{bmatrix} 0 & -i \\ i & 0 \end{bmatrix}$$

$$= \begin{bmatrix} 0 & i \\ i & 0 \end{bmatrix} - \begin{bmatrix} 0 & -i \\ -i & 0 \end{bmatrix}$$

$$= 2i\begin{bmatrix} 0 & 1 \\ 1 & 0 \end{bmatrix} \equiv 2i\sigma_x$$

$$(\bar{\sigma} \times \bar{\sigma})_y \overset{?}{=} 2i\sigma_y$$

$$= \sigma_z\sigma_x - \sigma_x\sigma_z$$

$$= \begin{bmatrix} 1 & 0 \\ 0 & -1 \end{bmatrix}\begin{bmatrix} 0 & 1 \\ 1 & 0 \end{bmatrix} - \begin{bmatrix} 0 & 1 \\ 1 & 0 \end{bmatrix}\begin{bmatrix} 1 & 0 \\ 0 & -1 \end{bmatrix}$$

$$= \begin{bmatrix} 0 & 1 \\ -1 & 0 \end{bmatrix} - \begin{bmatrix} 0 & -1 \\ 1 & 0 \end{bmatrix}$$

$$= 2\begin{bmatrix} 0 & 1 \\ -1 & 0 \end{bmatrix}$$

$$= 2i\begin{bmatrix} 0 & -i \\ i & 0 \end{bmatrix} \equiv 2i\sigma_y$$

$$(\bar{\sigma} \times \bar{\sigma})_z \overset{?}{=} 2i\sigma_z$$

$$= \sigma_x\sigma_y - \sigma_y\sigma_x$$

$$= \begin{bmatrix} 0 & 1 \\ 1 & 0 \end{bmatrix}\begin{bmatrix} 0 & -i \\ i & 0 \end{bmatrix} - \begin{bmatrix} 0 & -i \\ i & 0 \end{bmatrix}\begin{bmatrix} 0 & 1 \\ 1 & 0 \end{bmatrix}$$

$$= \begin{bmatrix} i & 0 \\ 0 & -i \end{bmatrix} - \begin{bmatrix} -i & 0 \\ 0 & i \end{bmatrix}$$

$$= 2i\begin{bmatrix} 1 & 0 \\ 0 & -1 \end{bmatrix} \equiv 2i\sigma_z$$

$$\boxed{\therefore \bar{S} \times \bar{S} = i\hbar\bar{S} \text{ and } \bar{S} \text{ is a valid angular momentum operator.}}$$

(b)

$$\chi_\uparrow = \begin{bmatrix} 1 \\ 0 \end{bmatrix}, \quad \chi_\downarrow = \begin{bmatrix} 0 \\ 1 \end{bmatrix}$$

$$S_z\chi_\uparrow = \frac{\hbar}{2}\sigma_z\chi_\uparrow$$

$$= \frac{\hbar}{2}\begin{bmatrix} 1 & 0 \\ 0 & -1 \end{bmatrix}\begin{bmatrix} 1 \\ 0 \end{bmatrix}$$

$$= \frac{\hbar}{2}\begin{bmatrix} 1 \\ 0 \end{bmatrix}$$

$$= \frac{\hbar}{2}\chi_\uparrow$$

$$\boxed{\therefore S_z\chi_\uparrow = \frac{\hbar}{2}\chi_\uparrow}$$

Similarly,

$$S_z \chi_\downarrow = \frac{\hbar}{2} \sigma_2 \chi_\downarrow$$

$$= \frac{\hbar}{2} \begin{bmatrix} 1 & 0 \\ 0 & -1 \end{bmatrix} \begin{bmatrix} 0 \\ 1 \end{bmatrix}$$

$$= \frac{\hbar}{2} \begin{bmatrix} 0 \\ -1 \end{bmatrix}$$

$$= -\frac{\hbar}{2} \begin{bmatrix} 0 \\ 1 \end{bmatrix}$$

$$= -\frac{\hbar}{2} \chi_\downarrow$$

$$\boxed{\therefore S_z \chi_\downarrow = -\frac{\hbar}{2} \chi_\downarrow}$$

(c)

$$\langle S^2 \rangle_\downarrow = \chi_\downarrow^T S^2 \chi_\downarrow$$

$$= \chi_\downarrow^T \left(S_x^2 + S_y^2 + S_z^2 \right) \chi_\downarrow$$

$$= \frac{\hbar^2}{4} \chi_\downarrow^T \left(\sigma_x^2 + \sigma_y^2 + \sigma_z^2 \right) \chi_\downarrow$$

But,

$$\sigma_x^2 = \begin{bmatrix} 0 & 1 \\ 1 & 0 \end{bmatrix} \begin{bmatrix} 0 & 1 \\ 1 & 0 \end{bmatrix}$$

$$= \begin{bmatrix} 1 & 0 \\ 0 & 1 \end{bmatrix}$$

$$= I$$

$$\sigma_y^2 = \begin{bmatrix} 0 & -i \\ i & 0 \end{bmatrix} \begin{bmatrix} 0 & -i \\ i & 0 \end{bmatrix}$$

$$= \begin{bmatrix} 1 & 0 \\ 0 & 1 \end{bmatrix}$$

$$= I$$

$$\sigma_z^2 = \begin{bmatrix} 1 & 0 \\ 0 & -1 \end{bmatrix} \begin{bmatrix} 1 & 0 \\ 0 & -1 \end{bmatrix}$$

$$= \begin{bmatrix} 1 & 0 \\ 0 & 1 \end{bmatrix}$$

$$= I$$

$$\therefore \langle S^2 \rangle_\downarrow = \frac{\hbar^2}{4} 3 \chi_\downarrow^T I \chi_\downarrow$$

$$= \frac{3\hbar^2}{4} [0 \ 1] \begin{bmatrix} 0 \\ 1 \end{bmatrix}$$

$$\boxed{\langle S^2 \rangle_\downarrow = \frac{3}{4} \hbar^2}$$

$$\langle S_z \rangle_\downarrow = \chi_\downarrow^T S_z \chi_\downarrow$$

$$= \frac{\hbar}{2} [1 \ 0] \begin{bmatrix} 1 & 0 \\ 0 & -1 \end{bmatrix} \begin{bmatrix} 0 \\ 1 \end{bmatrix}$$

$$= \frac{\hbar}{2} [0 \ 1] \begin{bmatrix} 0 \\ -1 \end{bmatrix}$$

$$\boxed{\langle S_z \rangle_\downarrow = -\frac{\hbar}{2}}$$

Similarly,

$$\langle S^2 \rangle_\uparrow = \chi_\uparrow^T S^2 \chi_\uparrow$$

$$= \frac{3\hbar^2}{4} [1 \ 0] \begin{bmatrix} 1 \\ 0 \end{bmatrix}$$

$$\boxed{\langle S^2 \rangle_\uparrow = +\frac{3\hbar^2}{4}}$$

$$\langle S_z \rangle_\uparrow = \chi_\uparrow^T S_z \chi_\uparrow$$

$$= \frac{\hbar}{2} [1 \ 0] \begin{bmatrix} 1 & 0 \\ 0 & -1 \end{bmatrix} \begin{bmatrix} 1 \\ 0 \end{bmatrix}$$

$$= \frac{\hbar}{2} [1 \ 0] \begin{bmatrix} 1 \\ 0 \end{bmatrix}$$

$$\boxed{\langle S_z \rangle_\uparrow = +\frac{\hbar}{2}}$$

1.3   Hund's Rules

When an atom is in the ground state, *Hund's rules* state that the electrons will occupy the orbitals so that S takes on its maximum possible value and L takes on the maximum possible value for this S. The total angular momentum is then[3]

---

[3] See, for example, Ashcroft and Mermin [8] and Martin [9].

- $J = |L - S|$ when the shell is less than half full, and
- $J = L + S$ when the shell is more than half full.

The ground states of atoms and ions are indicated with the notation $^{2S+1}X_J$, where $2S + 1$ is the number of states with a given S (called the multiplicity) and X is a letter corresponding to the value of L according to the convention shown in Table 1.1.

Hund's Rules can be visualized with the aid of a diagram with two rows as shown in Table 1.2. Values of m_l are listed in the first row, and values of m_s on the second row. The values of m_l are listed in decreasing order, starting with the largest positive value to the most negative value. This sequence is then repeated so that the diagram is divided into two sections, each with the complete range of m_l values. We then imagine adding electrons to the shell by making entries in the second row to represent the spin of the electron. To ensure that spin is maximized, all of the spins will be parallel until the shell is half full, then additional electrons must go into orbitals with spin opposite to the first half of the shell. To ensure that L is maximized subject to the value of spin, we enter electrons starting with the left-most entry in the m_s row with, say, spin up, and fill the shell up by making entries from left to right. For the left half of the diagram the shell is less than half full, so $J = |L - S|$ if the complete configuration is contained in the left half of the diagram, and $J = L + S$ if the configuration extends into the right side.

To illustrate how the diagram works, consider the case of Fe^{2+}. The electronic configuration can be found from a periodic table of the elements, and is $3d^6$. For the d shell $l = 2$ and so $m_l = 2, 1, 0, -1, -2$ as shown in Table 1.2. The six electrons are added beginning in the first position of the second row, filling the spin-up side with the sixth electron going into a spin-down orbital. Summing the m_l values gives $L = 2$, and summing the m_s values gives $S = 2$. The shell is more than half full, so $J = L + S = 4$. The spectroscopic notation for the state is then $^{2S+1}X_J = {}^5D_4$. In some materials, the incomplete shells are exposed to the electric fields from nearby ions, referred to as *crystal fields*. In such cases the interaction with the neighboring fixed ions destroys, or quenches, the orbital angular momentum of the incomplete shell. In such cases, the Hund rule ground state is obtained by setting $L = 0$. In other cases, the incomplete shells are shielded from the crystal fields by outer filled subshells, so that the ground states are unaffected by the neighboring ions, and Hund's rules can be directly applied.

Table 1.1 Symbols used to denote total orbital angular momentum in the Hund rule ground state

L	Symbol
0	S
1	P
2	D
3	F
4	G
5	H
6	I

Table 1.2 Hund rule ground state for Fe^{2+}. (Adapted from [7] with permission)

m_l :	2	1	0	-1	-2	2	1	0	-1	-2
m_s :	↑	↑	↑	↑	↑	↓				

Table 1.3 Hund rule ground state for Ce^{3+}

m_l :	3	2	1	0	-1	-2	-3	3	2	1	0	-1	-2	-3
m_s :	↑													

Problem 1.6 Rare-earth ions tend to first lose their $6s$ and $5d$ electrons, then their $4f$ electrons. The incomplete $4f$ subshell is shielded from the crystal fields by the outer $5s$ and $5p$ subshells so that the orbital angular momentum in these ions is not quenched, and Hund's Rules can be directly applied. Calculate the Hund rule ground states of the following ions and express the states in the spectroscopic notation $^{2S+1}X_J$: Ce^{3+}, Pr^{3+}, Nd^{3+}, Sm^{3+}, Eu^{3+}, Gd^{3+}, Tb^{3+}, Er^{3+}, and Lu^{3+}. Also calculate the Landé g factor for each state.

Solution 1.6

The electronic configuration of Ce^{3+} can be found from a periodic table of the elements, and is $4f^1$. As shown in Table 1.3, the single electron goes in the first position of the second row, giving $L = 3$ and $S = 1/2$.

Since the shell is less than half full, we have $J = |L - S| = 5/2$. Thus, we obtain ground state:

$$\boxed{^{2S+1}X_J = {}^2F_{5/2}.}$$

The g factor is calculated as

$$\begin{aligned} g &= \frac{3}{2} + \frac{S(S+1) - L(L+1)}{2J(J+1)} \\ &= \frac{3}{2} + \frac{\frac{1}{2}\left(\frac{3}{2}\right) - 3(4)}{2\left(\frac{5}{2}\right)\left(\frac{7}{2}\right)} \end{aligned} \tag{1.37}$$

Thus, $\boxed{g = 0.857.}$

Following this procedure, we can calculate the spectroscopic notation and g for the different ions, as shown in Table 1.4.

Table 1.4 Application of Hund's rules for different ions. Note that the $5d$ and $6s$ electrons in Gd^{3+}, Tb^{3+}, Eu^{3+}, and Lu^{3+} ions are ionized first

		m_l :	3	2	1	0	-1	-2	-3	3	2	1	0	-1	-2	-3	L	S	J	$^{2S+1}X_J$	g
Ce^{3+}	$4f^1$	m_s :	↑														3	$\frac{1}{2}$	$\frac{5}{2}$	$^2F_{5/2}$	0.86
Pr^{3+}	$4f^2$	m_s :	↑	↑													5	1	4	$^3H_4$	0.80
Nd^{3+}	$4f^3$	m_s :	↑	↑	↑												6	$\frac{3}{2}$	$\frac{9}{2}$	$^4I_{9/2}$	0.73
Sm^{3+}	$4f^5$	m_s :	↑	↑	↑	↑	↑										5	$\frac{5}{2}$	$\frac{5}{2}$	$^6H_{5/2}$	0.29
Eu^{3+}	$4f^6$	m_s :	↑	↑	↑	↑	↑	↑									3	3	0	$^7F_0$	NA
Gd^{3+}	$4f^7$	m_s :	↑	↑	↑	↑	↑	↑	↑								0	$\frac{7}{2}$	$\frac{7}{2}$	$^8S_{7/2}$	2
Tb^{3+}	$4f^8$	m_s :	↑	↑	↑	↑	↑	↑	↑	↓							3	3	6	$^7F_6$	1.5
Er^{3+}	$4f^{11}$	m_s :	↑	↑	↑	↑	↑	↑	↑	↓	↓	↓	↓				6	$\frac{3}{2}$	$\frac{15}{2}$	$^4I_{15/2}$	1.2
Lu^{3+}	$4f^{14}$	m_s :	↑	↑	↑	↑	↑	↑	↑	↓	↓	↓	↓	↓	↓	↓	0	0	0	$^1S_0$	NA

Problem 1.7 Iron group ions lose their $4s$ electrons before losing $3d$ electrons. Unlike the Rare Earths, there are no outer filled subshells to shield the $3d$ orbitals from the crystal fields of a host crystal. As a consequence, the orbital angular momentum of these ions is usually quenched in solids. Assuming that the orbital angular momentum is quenched, calculate the Hund rule ground states of the following ions and express the states in the spectroscopic notation $^{2S+1}X_J$: Cr^{2+}, Mn^{3+}, Fe^{3+}, Fe^{4+}, Co^{2+}, Ni^{2+}, Cu^{2+} and Cu^{3+}. Also calculate the Landé g factor for each state.

Solution 1.7
The valence electron configuration for Cr^{2+} is $3d^4$. Once again, following Hund's rules, we fill the orbitals as shown below:

m_l :	2	1	0	-1	-2	2	1	0	-1	-2
m_s :	↑	↑	↑	↑						

$L = 2$, but since orbital angular momentum is quenched, we take $L = 0$. In this case we have

$$S = 2, \quad J = S = 2 \Rightarrow \boxed{^{2S+1}X_J \equiv {}^5S_2}$$

$$\boxed{g = \frac{3}{2} + \frac{2(3)}{2(2)3} = 2}$$

Table 1.5 shows the calculation for all the other ions.

Problem 1.8 The magnetic oxide yttrium iron garnet $Y_3Fe_5O_{12}$ is often used in microwave devices. This material has a complicated cubic crystal structure with eight formula units per cell. The five Fe^{3+} ions per formula unit are distributed

Table 1.5 Application of Hund's rules for different ions. We assume fully quenched orbital angular momentum and set $L = 0$ while calculating $J = L + S$

		m_l :	2	1	0	-1	-2	2	1	0	-1	-2	L	S	J	$^{2S+1}X_J$	g
Cr^{2+}	$3d^4$	m_s :	↑	↑	↑	↑							2	2	2	$^5S_2$	2
Mn^{3+}	$3d^4$	m_s :	↑	↑	↑	↑							2	2	2	$^5S_2$	2
Fe^{3+}	$3d^5$	m_s :	↑	↑	↑	↑	↑						0	$\frac{5}{2}$	$\frac{5}{2}$	$^6S_{5/2}$	2
Fe^{4+}	$3d^4$	m_s :	↑	↑	↑	↑							2	2	2	$^5S_2$	2
Co^{2+}	$3d^7$	m_s :	↑	↑	↑	↑	↑	↓	↓				3	$\frac{3}{2}$	$\frac{3}{2}$	$^4S_{3/2}$	2
Ni^{2+}	$3d^8$	m_s :	↑	↑	↑	↑	↑	↓	↓	↓			3	1	3	$^3S_1$	2
Cu^{2+}	$3d^9$	m_s :	↑	↑	↑	↑	↑	↓	↓	↓	↓		2	$\frac{1}{2}$	2	$^2S_{\frac{1}{2}}$	2
Cu^{3+}	$3d^8$	m_s :	↑	↑	↑	↑	↑	↓	↓	↓			3	1	3	$^3S_1$	2

between antiparallel sublattices, giving the material a ferrimagnetic structure. The net difference is one Fe^{3+} ion per formula unit. The edge of the cubic cell is 12.38 A°. Calculate the maximum value of the magnetization M, i.e., its value at $T = 0$ K.

Solution 1.8 For Fe^{3+}

$$J = \frac{5}{2}, \ \mu = \mu_B \frac{5}{2}$$

where the *Bohr magneton* is defined as [7],

$$\mu_B = \frac{|q|\hbar}{2m_q} = 9.27 \left(10^{-24}\right) \frac{J}{T}$$

$$\frac{\mu}{V_c} = g \frac{8\mu_B \frac{5}{2}}{\left[12.38(10^{-10})\right]^3} = 2\frac{20\mu_B}{1.90(10^{-27})} = 2\frac{20(9.27)(10^{-24})}{1.90(10^{-27})}.$$

Hence,

$$\boxed{M_s(0) = 195.5 \text{ kA/m.}}$$

References

1. R.P. Feynman, R.B. Leighton, M. Sands, *The Feynman Lectures on Physics*, vol. I. (Addison-Wesley, Reading, 1964)
2. H. Goldstein, C.P. Poole, J.L. Safko, *Classical Mechanics*, 3rd edn. (Addison-Wesley, Cambridge, 2001)
3. R.M. Eisberg, *Fundamentals of Modern Physics* (Wiley, New York, 1961)
4. R.P. Feynman, R.B. Leighton, M. Sands, *The Feynman Lectures on Physics*, vol. III. (Addison-Wesley, Reading, 1964)
5. E. Merzbacher, *Quantum Mechanics*, 3rd edn. (Wiley, New York, 1998)
6. L.I. Schiff, *Quantum Mechanics* (McGraw-Hill, New York, 1968)
7. D.D. Stancil, A. Prabhakar, *Spin Waves: Theory and Applications* (Springer, New York, 2009)
8. N. Ashcroft, N.D. Mermin, *Solid State Physics* (Holt, Rinehart and Winston, New York, 1976)
9. D.H. Martin, *Magnetism in Solids* (M.I.T. Press, Cambridge, 1967)

Quantum Theory of Spin Waves

2

When the spacing between spins is small enough that the wave functions of the particles overlap, spins are coupled to each other through a quantum mechanical mechanism called the exchange interaction. In the simplest treatment of this interaction, spins are coupled only to their nearest neighbors, giving rise to what is known as the Heisenberg Hamiltonian. The ground and excited states of molecular hydrogen are explored in Problem 2.1 as a way of gaining some insight into the origin of the exchange interaction.

The quantum mechanical harmonic oscillator and its associated raising and lowering operators in the occupation number (or second quantization) formalism are important to many phenomena in physics, including spin waves in particular. The harmonic oscillator is considered in Problems 2.2 and 2.3. Building on the insights from the harmonic oscillator problem, spin raising and lowering operators are discussed in Problem 2.4. Spin wave excitations on a linear chain are considered in Problems 2.5 and 2.6.

2.1   Hamiltonians for Coupled Spins

As we saw in the last chapter, spins placed in a magnetic field precess around the direction of the applied field. When treated quantum mechanically, this is referred to as *Larmor precession*. Spin waves can be thought of as a sinusoidal variation in the phase of this precession from spin to spin. Charged particles with spin (such as electrons) interact through their dipolar magnetic fields, but when the spacing between spins is small enough that the *wave functions* of the particles overlap, a quantum mechanical interaction dominates that arises from electrostatic forces along with the Pauli Exclusion principle. This is referred to as the *exchange interaction*, because when the energy is calculated using the wave functions and Schrodinger's equation,

© The Author(s), under exclusive license to Springer Nature Switzerland AG 2021
D. D. Stancil and A. Prabhakar, *Spin Waves*,
https://doi.org/10.1007/978-3-030-68582-9_2

a key term in the interaction can be interpreted as representing the probability that the spins will exchange places.

The simplest system that provides some understanding of the origin of this interaction is molecular hydrogen.[1] Approximate 2-electron wave functions ψ can be constructed by the product between a spatial function ψ and a spin function χ. According to the statistics of fermions, interchanging the two electrons should change the sign of the overall wave function. We can construct wave functions with overall antisymmetric behavior from the products of symmetric and antisymmetric spatial functions and spin state vectors, $\psi_{A,S}$, $\chi_{A,S}$, where A and S indicate antisymmetric and symmetric functions, respectively.

The solution to the molecular hydrogen problem yields two energy levels. The higher energy level is obtained for three different spin configurations, and is called a triplet state. Each of the triplet states has the form of an antisymmetric spatial function multiplied by a symmetric spin state vector: $\psi_A \chi_S$. The lower energy level is taken to be the ground state and occurs for a single spin configuration. Consequently, it is referred to as a singlet state and is of the form of a symmetric spatial function and an antisymmetric spin state vector: $\psi_S \chi_A$. The antisymmetric spin state vector represents a total spin of 0, while the symmetric spin state vector represents a total spin of 1.

Using these properties, we can construct an effective *Hamiltonian* that gives the correct energy when applied to a specific wave function. For example,

$$\mathcal{H}^{\text{spin}} = \mathcal{E}_S + \frac{1}{2\hbar^2} [\mathcal{E}_T - \mathcal{E}_S] S^2, \tag{2.1}$$

where $\mathcal{E}_T$ and $\mathcal{E}_S$ are the energies of the triplet and singlet states, respectively, and S^2 is the square of the spin operator with the property (see Chap. 1):

$$S^2 \psi = \hbar^2 s(s+1) \psi. \tag{2.2}$$

Here s is the quantum number for total spin associated with the state ψ.

Another form of a Hamiltonian that will give correct expectation values by sensing only the spin can be obtained by expanding the operator for the square of the total spin: $S^2 = (\mathbf{S}_1 + \mathbf{S}_2) \cdot (\mathbf{S}_1 + \mathbf{S}_2) = S_1^2 + S_2^2 + 2\mathbf{S}_1 \cdot \mathbf{S}_2$. The last term is clearly dependent on the relative spin configurations at sites 1 and 2. Using this expansion, the spin Hamiltonian can be written as

$$\mathcal{H}^{\text{spin}} = \mathcal{E}_S + \frac{1}{2\hbar^2} [\mathcal{E}_T - \mathcal{E}_S] \left(S_1^2 + S_2^2 + 2\mathbf{S}_1 \cdot \mathbf{S}_2 \right). \tag{2.3}$$

---

[1]The exchange interaction introduced through the energy states of molecular hydrogen is discussed in Mattis [1]. Treatments for exchange can also be found in Martin [2] and Rado and Suhl [3].

Since both electrons are spin 1/2 particles, we have $S_i^2 \chi = \hbar^2 s_i (s_i + 1) \chi = (3\hbar^2/4)\chi$. The spin Hamiltonian can be further simplified with this result:

$$\mathcal{H}^{\text{spin}} = \mathcal{E}_S + \frac{1}{2\hbar^2} [\mathcal{E}_T - \mathcal{E}_S] \left(\frac{3\hbar^2}{2} + 2\mathbf{S}_1 \cdot \mathbf{S}_2 \right)$$

$$= \frac{1}{4} (\mathcal{E}_S + 3\mathcal{E}_T) + \frac{1}{\hbar^2} [\mathcal{E}_T - \mathcal{E}_S] \mathbf{S}_1 \cdot \mathbf{S}_2. \tag{2.4}$$

Solving for the term that depends on spin orientation, we obtain

$$\frac{1}{\hbar^2} [\mathcal{E}_T - \mathcal{E}_S] \mathbf{S}_1 \cdot \mathbf{S}_2 \equiv -2\frac{J}{\hbar^2} \mathbf{S}_1 \cdot \mathbf{S}_2 = \mathcal{H}^{\text{spin}} - \frac{1}{4} (\mathcal{E}_S + 3\mathcal{E}_T). \tag{2.5}$$

Here we have introduced the constant J to represent the strength of the spin-dependent coupling between the particles, which is related to the exchange interaction. This form of the Hamiltonian explicitly showing the energy dependence on the dot product between two spins (or spin operators) is referred to as the *Heisenberg Hamiltonian*. Equation (2.5) will be useful for the calculations in Problem 2.1.

A third form of a Hamiltonian that gives the expected energy by only sensing the spin can be constructed using the *Pauli spin exchange operator P*. This operator simply interchanges the two spins, so that

$$P\chi_S = \chi_S, \quad P\chi_A = -\chi_A. \tag{2.6}$$

Using this operator, a Hamiltonian equivalent to the Heisenberg Hamiltonian can be written

$$\mathcal{H}^P = -\frac{J}{2}(2P - 1). \tag{2.7}$$

Problem 2.1 In this problem we will verify the equivalence of the Heisenberg Hamiltonian and the alternative form expressed in terms of the Pauli spin exchange operator when applied to the hydrogen triplet state ψ_{AXS}.

(a) Calculate the expected value of the Heisenberg Hamiltonian by evaluating

$$-2\frac{J}{\hbar^2} \mathbf{S}_1 \cdot \mathbf{S}_2 \, \psi_{AXS} = \left(\mathcal{H}^{\text{spin}} - \frac{1}{4} [\mathcal{E}_S + 3\mathcal{E}_T] \right) \psi_{AXS}, \tag{2.8}$$

where $J = (\mathcal{E}_S - \mathcal{E}_T)/2$.

(b) Calculate the expected value of the alternative Hamiltonian by evaluating

$$\mathcal{H}^P \psi_{AXS} = -\frac{J}{2}(2P - 1) \, \psi_{AXS}. \tag{2.9}$$

Solution 2.1

(a) The expectation value of the Heisenberg Hamiltonian can be calculated as follows:

$$-\frac{2J}{\hbar^2}\bar{S}_1 \cdot \bar{S}_2 \psi_{A\chi S} = \left(H^{\text{spin}} - \frac{1}{4}[\mathcal{E}_S + 3\mathcal{E}_T] \right) \psi_{A\chi S}$$

$$= \left(\mathcal{E}_T - \frac{1}{4}[\mathcal{E}_S + 3\mathcal{E}_T] \right) \psi_{A\chi S}$$

$$= \left(\frac{1}{4}\mathcal{E}_T - \frac{1}{4}\mathcal{E}_S \right) \psi_{A\chi S}$$

$$-\frac{2J}{\hbar^2}\bar{S}_1 \cdot \bar{S}_2 \psi_{A\chi S} = \frac{1}{4}\left(\mathcal{E}_T - \mathcal{E}_S \right) \psi_{A\chi S}$$

$$\boxed{\therefore -\frac{2J}{\hbar^2}\langle \bar{S}_1 \cdot \bar{S}_2 \rangle = \frac{1}{4}\left(\mathcal{E}_T - \mathcal{E}_S \right)}$$

(b) The expectation value of the alternative Hamiltonian can be calculated in a similar way:

$$H^P \psi_{A\chi S} = -\frac{J}{2}\left(2P - 1 \right) \psi_{A\chi S}$$

$$= -\frac{J}{2}\psi_{A\chi S}$$

$$= -\frac{1}{4}\left(\mathcal{E}_S - \mathcal{E}_T \right) \psi_{A\chi S}$$

$$H^P \psi_{A\chi S} = \frac{1}{4}\left(\mathcal{E}_T - \mathcal{E}_S \right) \psi_{A\chi S}$$

$$\boxed{\therefore \langle H^P \rangle = \frac{1}{4}\left(\mathcal{E}_T - \mathcal{E}_S \right)}$$

We thus establish the equivalence between the two forms of the Hamiltonian.

2.2   Harmonic Oscillator

An important problem in quantum mechanics is that of a charged particle in a quadratic potential well. This is the quantum analog of a classical spring–mass system, and is referred to as a harmonic oscillator.[2] In the case of the classical system, the potential energy increases as the square of the displacement of the spring from equilibrium, and in the quantum mechanical problem the potential energy of the

---

[2] A detailed discussion of the quantum mechanical harmonic oscillator can be found in most introductory texts on quantum mechanics, e.g.., Merzbacher [4], Schiff [5] or Shankar [6].

charged particle also increases with the square of the distance from the center equilibrium position. Classically, however, the spring–mass system can have any energy, but the energy levels of the quantum mechanical harmonic oscillator are quantized.

The time-independent form of Schrodinger's equation for the harmonic oscillator is given by

$$-\frac{\hbar^2}{2m}\frac{d^2\psi}{dx^2} + \frac{1}{2}m\omega^2 x^2 \psi = \mathcal{E}\psi, \tag{2.10}$$

where m is the mass of the particle, and ω is the natural frequency of the oscillator. The solutions to the harmonic oscillator have energies

$$\mathcal{E}_n = \hbar\omega\left(n + \frac{1}{2}\right), \tag{2.11}$$

and wave functions

$$\psi_n(x) = \frac{e^{-\frac{x^2}{2\alpha^2}} H_n(x/\alpha)}{\sqrt{\alpha 2^n n! \sqrt{\pi}}}, \tag{2.12}$$

where $\alpha = \sqrt{\hbar/m\omega}$. Here $H_n(\xi)$ are the *Hermite polynomials*. The first few polynomials are

$$H_0(\xi) = 1,$$
$$H_1(\xi) = 2\xi, \tag{2.13}$$
$$H_2(\xi) = -2 + 4\xi^2.$$

Problem 2.2 Plot the harmonic oscillator wave function (2.12) for $n = 1, 2, 3$, for $\alpha = 1$.

Solution 2.2

A Python program to calculate and plot the first three harmonic oscillator wave functions is given below (Fig. 2.1).

```
from matplotlib import pyplot as plt
import numpy as np

# create x array
x = np.arange(-5,5,0.01)

# Gaussian factor
G = np.exp(-x**2/2)

# n=0
psi0 = G/np.pi**(1/4)
```

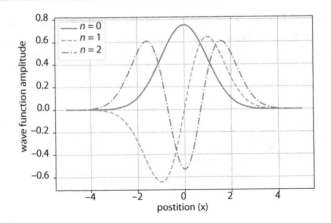

Fig. 2.1 Harmonic oscillator wave functions

```
# n=1
H1 = 2*x
C1 = (2*np.pi**(1/2))**(1/2)
psi1 = G*H1/C1

# n=2
H2 = -2+4*x**2
C2 = (2**2*2*np.pi**(1/2))**(1/2)
psi2 = G*H2/C2

# plot and download
fig, ax=plt.subplots()
ax.plot(x,psi0,label='$n=0$')
ax.plot(x,psi1,linestyle='--',label='$n=1$')
ax.plot(x,psi2,linestyle='-.',label='$n=2$')
ax.legend(loc='upper left')
ax.set(xlabel='position (x)', ylabel='wave function amplitude')
ax.grid()
fig1 = plt.gcf()
plt.show()}
```

2.3   Raising and Lowering Operators

Suppose we have operators that have the following effects on the nth state:

$$a\psi_n = \sqrt{n}\psi_{n-1}, \quad a^\dagger\psi_n = \sqrt{n+1}\psi_{n+1}. \tag{2.14}$$

Since the effect of a is to take the state ψ_n into the state ψ_{n-1}, it is called a *lowering* operator. Similarly, since the effect of $a^\dagger$ is to take the state ψ_n into the state ψ_{n+1}, it is referred to as a *raising* operator. For the harmonic oscillator wave functions, raising and lowering operators are given explicitly by

$$a^\dagger = \frac{1}{\sqrt{2}}\left(\frac{x}{\alpha} - \alpha\frac{d}{dx}\right),\qquad(2.15)$$

$$a = \frac{1}{\sqrt{2}}\left(\frac{x}{\alpha} + \alpha\frac{d}{dx}\right).\qquad(2.16)$$

Problem 2.3 Show that

$$a^\dagger \psi_n = \sqrt{n+1}\,\psi_{n+1}\qquad(2.17)$$

where

$$a^\dagger = \frac{1}{\sqrt{2}}\left(\frac{x}{\alpha} - \alpha\frac{d}{dx}\right)\qquad(2.18)$$

and

$$\psi_n(x) = \frac{e^{-\frac{x^2}{2\alpha^2}}\,H_n\left(\frac{x}{\alpha}\right)}{\sqrt{\alpha 2^n n!\sqrt{\pi}}}.\qquad(2.19)$$

Solution 2.3

$$a^+\psi_n \overset{?}{=} \sqrt{n+1}\psi_{n+1}$$

$$a^+\psi_n = \frac{1}{\sqrt{2}}\left(\frac{x}{\alpha} - \alpha\frac{d}{dx}\right)\psi_n$$

The harmonic oscillator wave functions, described in [7, Sec. 2.6], have the property

$$\frac{d}{dx}\psi_n = \frac{1}{\alpha}\left(\psi_{n-1}\sqrt{\frac{n}{2}} - \psi_{n+1}\sqrt{\frac{n+1}{2}}\right)$$

$$\Rightarrow a^+\psi_n = \frac{1}{\sqrt{2}}\left[\frac{x}{\alpha}\psi_n - \psi_{n-1}\sqrt{\frac{n}{2}} + \psi_{n+1}\sqrt{\frac{n+1}{2}}\right].\qquad(2.20)$$

Consider

$$
\frac{x}{\alpha}\psi_n - \psi_{n-1}\sqrt{\frac{n}{2}} = \frac{x\, e^{\frac{-x^2}{2\alpha^2}} H_n\left(\frac{x}{\alpha}\right)}{\alpha\,\sqrt{\alpha 2^n n!\sqrt{\pi}}} - \sqrt{\frac{n}{2}}\,\frac{e^{\frac{-x^2}{2\alpha^2}} H_{n-1}\left(\frac{x}{\alpha}\right)}{\sqrt{\alpha 2^{n-1}(n-1)!\sqrt{\pi}}}
$$

$$
= e^{\frac{-x^2}{2\alpha^2}}\left[\frac{x}{\alpha}\frac{H_n}{\sqrt{\alpha 2^n n!\sqrt{\pi}}} - \frac{n H_{n-1}}{\sqrt{\alpha 2^n n!\sqrt{\pi}}}\right]
$$

$$
= \frac{e^{\frac{-x^2}{2\alpha^2}}}{\sqrt{\alpha 2^n n!\sqrt{\pi}}}\left[\frac{x}{\alpha}H_n - n H_{n-1}\right].
$$

The Hermite polynomials, described in [7, Sec. 2.6.1], have the *recurrence relation*

$$
\xi H_n - n H_{n-1} = \frac{1}{2}H_{n+1}
$$

$$
\Rightarrow \quad \frac{x}{\alpha}\psi_n - \psi_{n-1}\sqrt{\frac{n}{2}} = \frac{e^{\frac{-x^2}{2\alpha^2}}}{\sqrt{\alpha 2^n n!\sqrt{\pi}}}\frac{1}{2}H_{n+1}
$$

$$
= \frac{e^{\frac{-x^2}{2\alpha^2}}}{\sqrt{\alpha 2^{n+1}(n+1)!\sqrt{\pi}}}\sqrt{\frac{n+1}{2}}H_{n+1}
$$

$$
\therefore \quad \frac{x}{\alpha}\psi_n - \sqrt{\frac{n}{2}}\psi_{n-1} = \sqrt{\frac{n+1}{2}}\psi_{n+1}. \tag{2.21}
$$

Using (2.20) and (2.21) gives us

$$
a^+\psi_n = \frac{1}{\sqrt{2}}\left[2\sqrt{\frac{n+1}{2}}\psi_{n+1}\right]
$$

$$
\boxed{a^+\psi_n = \sqrt{n+1}\,\psi_{n+1}.}
$$

2.4   Bra-Ket Notation

In Problem 1.5 we represented the spinors χ as column vectors, and expectation values involved inner products of the form $\chi^T O \chi$, where χ^T is the matrix transpose of χ, and O is an operator represented by a matrix. Another notation that is widely used is referred to as "bra-ket" notation. In this notation, a "ket" refers to the notation $|a\rangle$, and a "bra" refers to the notation $\langle a|$, where a is some convenient label for the state. The bra is defined as the transpose complex conjugate of a ket: $\langle a| = (|a\rangle^T)^*$. The terminology is a play on words, since the inner product $\langle a|a\rangle$ forms a "bra-ket."

To relate this to our previous notation for a single spin, we can write

$$
|\uparrow\rangle = \chi_\uparrow = \begin{bmatrix}1\\0\end{bmatrix}, \quad |\downarrow\rangle = \chi_\downarrow = \begin{bmatrix}0\\1\end{bmatrix}, \tag{2.22}
$$

and

$$\langle\uparrow| = \chi_{\uparrow}^T = \begin{bmatrix} 1 & 0 \end{bmatrix}, \quad \langle\downarrow| = \chi_{\downarrow}^T = \begin{bmatrix} 0 & 1 \end{bmatrix}. \tag{2.23}$$

Since the vectors are real, the complex conjugate operation leaves the vectors unchanged in this case.

Instead of using the labels $\uparrow$ and $\downarrow$, a label with equivalent information for spin 1/2, but also useful for systems with larger spin is $|s_z\rangle$, where s_z is the z component of angular momentum. As another example, in Sec. 2.6 on Spin Waves we will have occasion to use $|n\rangle$ to represent the state of a linear chain of spins in which the nth spin is flipped.

2.5   Spin Raising and Lowering Operators

In Chap. 1 we found that $L_{\pm}\psi_m \sim \psi_{m\pm1}$, but we did not specifically address what the constant of proportionality might be. Revisiting this in the context of spin angular momentum and using the bra-ket notation, we can write, for example,

$$S_j^- |s_{jz}\rangle = \lambda |s_{jz} - 1\rangle, \tag{2.24}$$

where the operator

$$S_j^- = S_{jx} - iS_{jy}$$

is the spin lowering operator for the jth site.

To find $|\lambda|$, we can evaluate the expectation

$$\langle s_{jz}| S_j^+ S_j^- |s_{jz}\rangle = |\lambda|^2. \tag{2.25}$$

Problem 2.4 In this problem we will calculate the constant of proportionality λ discussed above.

(a) As a first step in calculating the expectation $\langle s_{jz}| S_j^+ S_j^- |s_{jz}\rangle$, show that

$$S_j^2 = \frac{1}{2}\left(S_j^+ S_j^- + S_j^- S_j^+\right) + S_{jz}^2. \tag{2.26}$$

(b) In the above expression, solve for $S_j^+ S_j^-$ and calculate its expectation value to show that

$$S_j^- |s_{jz}\rangle = \hbar \left(s(s+1) - s_{jz}\left(s_{jz} - 1\right)\right)^{1/2} |s_{jz} - 1\rangle. \tag{2.27}$$

Here we assume an array of identical spins such that the total spin at each site is the same, i.e., $s_i = s_j = s$.

Solution 2.4

(a) To show that (2.26) is true, we substitute explicit expressions for S_j^+ and S_j^- and simplify:

$$S_j^2 = \frac{1}{2} \left((S_{jx} + i S_{jy})(S_{jx} - i S_{jy}) + (S_{jx} - i S_{jy})(S_{jx} + i S_{jy}) \right) + S_{jz}^2$$

$$= \frac{1}{2} \left(S_{jx}^2 + S_{jy}^2 + S_{jx}^2 + S_{jy}^2 \right) + S_{jz}^2$$

and $$\boxed{S_j^2 = S_{jx}^2 + S_{jy}^2 + S_{jz}^2 .}$$

(b) From the commutation relation $[S^+, S^-]$ (see Problem 1.2),

$$S_j^- S_j^+ = S_j^+ S_j^- - 2\hbar S_{jz}.$$

Substituting this result into (2.26) gives the following:

$$S_j^2 = \frac{1}{2} \left(2 S_j^+ S_j^- - 2\hbar S_{jz} \right) + S_{jz}^2$$

$$= S_j^+ S_j^- - \hbar S_{jz} + S_{jz}^2$$

$$= S_j^+ S_j^- + S_{jz} \left(S_{jz} - \hbar \right)$$

or

$$S_j^+ S_j^- = S_j^2 - S_{jz} \left(S_{jz} - \hbar \right).$$

So,

$$\left\langle s_{jz} \left| S_j^+ S_j^- \right| s_{jz} \right\rangle = \left\langle s_{jz} \left| S_j^2 - S_{jz} \left(S_{jz} - \hbar \right) \right| s_{jz} \right\rangle$$

$$= \hbar^2 s \left(s + 1 \right) - \hbar^2 s_{jz} \left(s_{jz} - 1 \right)$$

$$|\lambda|^2 = \hbar^2 \left[s \left(s + 1 \right) - s_{jz} \left(s_{jz} - 1 \right) \right]$$

$$\lambda = \hbar \left[s \left(s + 1 \right) - s_{jz} \left(s_{jz} - 1 \right) \right]^{\frac{1}{2}}$$

$$\therefore S_j^- \left| s_{jz} \right\rangle = \lambda \left| s_{jz} - 1 \right\rangle$$

and $$\boxed{S_j^- \left| s_{jz} \right\rangle = \hbar \left[s \left(s + 1 \right) - s_{jz} \left(s_{jz} - 1 \right) \right]^{\frac{1}{2}} \left| s_{jz} - 1 \right\rangle .}$$

2.6 Spin Waves

Suppose we consider a linear chain of N spins with ferromagnetic exchange interactions. In this case the ground state will be the state with all spins parallel. If we further apply a magnetic field, then the ground state will be the state with the spins also anti-parallel with the applied magnetic field (for negatively charged particles).

The first excited state will be one in which only 1 spin in the chain is flipped. In general a superposition of all possible states with 1 flipped spin can be written

$$|\psi\rangle = \sum_n |n\rangle \langle n|\psi\rangle, \qquad (2.28)$$

where $|n\rangle$ is the state in which the flipped spin is on the nth site, and $\langle n|\psi\rangle$ is the probability amplitude that the system is in the state with the flipped spin at the nth site. Now suppose that all of the coefficients have the same magnitude, but differ in phase, i.e., $\langle n|\psi\rangle = \exp(i\theta_n)/\sqrt{N}$. The factor $1/\sqrt{N}$ normalizes the basis states so that the probability of measuring the flipped spin on any site is the same and equal to $1/N$.

A *spin wave* consists of such a superposition state, but with a progressive phase shift between sites. This can be written

$$\psi(x_n) \equiv \langle n|\psi\rangle = \frac{e^{ikx_n}}{\sqrt{N}}, \qquad (2.29)$$

where $x_n = na$, and a is the lattice spacing between sites. We can visualize the spin wave as a phase shift of the spin precession that varies linearly with position as shown in Fig. 2.2.

For spin 1/2, the z component of the spin on a given site in the ground state is $s_z = 1/2$ (in units of $\hbar$), and in the excited state the spin is $s_z - 1 = -1/2$. The situation in which a single flipped spin is spread evenly over a chain of N spins suggests a classical picture where the average z component of spin on a given site is $s_z - 1/N$. This picture can be used to estimate the average angle that the spin precession makes with the z axis, or average "cone angle" for the spin wave. A spin wave consisting of a single flipped spin on the chain is also called a *magnon*.

Problem 2.5 Consider a linear chain of N spins. The exact solution on a finite chain will depend on the boundary conditions at the ends. A common choice in solid state physics is the periodic boundary condition, i.e., $\psi(x_1) = \psi(x_{1+N})$.

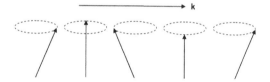

Fig. 2.2 A spin wave can be visualized as a linear shift in phase in the direction of propagation. (Reproduced from [7] with permission.)

(a) If $x_n = na$ and $\psi(x) = e^{ikx}$, show that the allowed values of k are $k_m = 2\pi m / Na$, where $m = 0, \pm 1, \pm 2 \ldots \pm N/2$ (assuming N is even).

(b) With the definitions from Part (a), show that

$$\sum_{j=0}^{N-1} e^{ik_m x_j} = 0, \quad k_m \neq 0 \tag{2.30}$$

$$\sum_{j=0}^{N-1} e^{ik_m x_j} = N, \quad k_m = 0. \tag{2.31}$$

Solution 2.5

(a) Imposing the periodic boundary condition gives

$$\psi(x_1) = \psi(x_{1+N}), \quad x_n = na, \quad \psi = e^{ikx}$$
$$\therefore e^{ika} = e^{ik(1+N)a}$$
$$1 = e^{ikNa}$$
$$\Rightarrow kNa = 2\pi m, \quad m = \text{integer},$$

$$\text{or,} \quad \boxed{k_m = \frac{2\pi m}{Na}}.$$

While m can be any integer, the values $|m| > \frac{N}{2}$ are equivalent to the values for $|m| < \frac{N}{2}$. For example (Fig. 2.3)

$$\exp(ik_{(N/2+1)}a) = \exp\left(i\frac{2\pi}{Na}\left(\frac{N}{2}+1\right)a\right)$$
$$= \exp\left(i\left(\pi + \frac{2\pi}{N}\right)\right)$$
$$= \exp\left(i\left(\pi + \frac{2\pi}{N}\right) - 2\pi i\right)$$
$$= \exp\left(i\left(-\pi + \frac{2\pi}{N}\right)\right)$$
$$= \exp\left(i\left(\frac{2\pi}{Na}\right)\left(-\frac{N}{2}+1\right)a\right)$$
$$= \exp\left(ik_{(-\frac{N}{2}+1)}a\right)$$

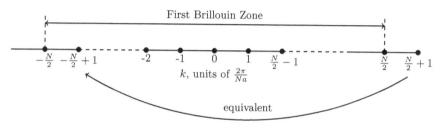

Fig. 2.3 $N/2 + 1$ is an equivalent point to $-N/2 + 1$

Hence, we conclude that $m = \frac{N}{2} + 1$ gives the same result as $m = -\frac{N}{2} + 1$. Thus, the distinct values of m are $m = 0, \pm 1, \pm 2 \cdots \pm \frac{N}{2}$ assuming N is even. Following a similar argument, it is also clear that the state with $m = \frac{N}{2}$ is equivalent to the state $m = -\frac{N}{2}$, so only one of the two need be included. The unique values of k_m are therefore $k_m = -\frac{N}{2}, -\frac{N}{2} + 1, -\frac{N}{2} + 2 \cdots \frac{N}{2} - 1$, or equivalently $k_m = -\frac{N}{2} + 1, -\frac{N}{2} + 2 \cdots \frac{N}{2} - 1, \frac{N}{2}$. Either of these ranges of k_m make up the first *Brillouin Zone*.

(b) If $k_m = 0$ then

$$\sum_{j=0}^{N-1} e^{ik_m x_j} = \sum_{j=0}^{N-1} x^0 = \sum_{j=0}^{N-1} 1 = N$$

For $k_m \neq 0$

$$\sum_{j=0}^{N-1} e^{ik_m x_j} = \sum_{j=0}^{N-1} e^{i\frac{2\pi m}{N} j} = \sum_{j=0}^{N-1} X^j$$

where

$$X = e^{i\frac{2\pi m}{N}}.$$

Now,

$$\sum_{j=0}^{N-1} X^j = 1 + X + X^2 \cdots X^{N-1}$$

and

$$X \sum_{j=0}^{N-1} X^j = X + X^2 + \cdots X^N.$$

Subtract to get:

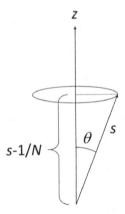

Fig. 2.4 Classical vector of length s precessing about the z-axis

$$X \sum_{j=0}^{N-1} X^j - \sum_{j=0}^{N-1} X^j = X^N - 1$$

$$(X - 1) \sum_{j=0}^{N-1} X^j = X^N - 1$$

$$\therefore \sum_{j=0}^{N-1} X^j = \frac{X^N - 1}{X - 1} = \frac{e^{i 2\pi m} - 1}{e^{i \frac{2\pi m}{N}} - 1} = \frac{1 - 1}{e^{i \frac{2\pi m}{N}} - 1}$$

Therefore,

$$\boxed{\sum_{j=0}^{N-1} e^{i k_m x_j} = 0, \ k_m \neq 0.}$$

Problem 2.6 Consider a classical vector of length s precessing about the z axis. If the z component of s is $s - 1/N$ with $s \ll N$, show that the cone angle of the precession (i.e., the angle of s with respect to the z axis) is approximately $\sqrt{2/(Ns)}$.

Solution 2.6

Consider a vector of length s making an angle θ with the z axis such that the projection along the z axis is $s - 1/N$ as shown in Fig. 2.4

$$\cos \theta = \frac{s - \frac{1}{N}}{s} = 1 - \frac{1}{Ns}$$

But,

$$\cos \theta \approx 1 - \frac{\theta^2}{2}$$

$$\therefore 1 - \frac{\theta^2}{2} \approx 1 - \frac{1}{Ns}$$

$$\theta^2 \approx \frac{2}{Ns}$$

$$\boxed{\theta \approx \sqrt{\frac{2}{Ns}}.}$$

References

1. D.C. Mattis, *The Theory of Magnetism I: Statics and Dynamics* (Springer-Verlag, Berlin, 1981)
2. D.H. Martin, *Magnetism in Solids* (M.I.T. Press, Cambridge, MA, 1967)
3. G.T. Rado, H. Suhl (eds.), *Magnetism I* (Academic Press, New York, 1963)
4. E. Merzbacher, *Quantum Mechanics*, 3rd edn. (John Wiley & Sons, New York, 1998)
5. L.I. Schiff, *Quantum Mechanics* (McGraw-Hill, New York, 1968)
6. R. Shankar, *Principles of Quantum Mechanics*, 2nd edn. (Springer, Boston, MA, 1994)
7. D.D. Stancil, A. Prabhakar, *Spin Waves: Theory and Applications* (Springer, New York, 2009)

Magnetic Susceptibilities

3

In macroscopic media, we are interested in the behavior of the average density of magnetic dipoles when an external magnetic field is applied. We are specifically interested in the net *magnetization* (magnetic dipole moment per unit volume) that exists either spontaneously, or is induced in response to an external magnetic field. When the relation between the induced magnetization and the applied field is linear, we express the magnetization in terms of a magnetic susceptibility tensor $\overline{\overline{\chi}}$ such that

$$\mathbf{M} = \mathbf{M}_0 + \overline{\overline{\chi}} \cdot \mathbf{H}. \tag{3.1}$$

For isotropic media, the susceptibility reduces to a scalar constant times the identity matrix, and $\mathbf{M}$ and $\mathbf{H}$ are collinear.

In this chapter we introduce the different types of magnetic materials (diamagnetic, paramagnetic, ferromagnetic, ferrimagnetic, and antiferromagnetic) and learn how to estimate their susceptibility. Problem 3.1 describes the use of the Langevin diamagnetism formula, while Problem 3.2 introduces the Brillouin function for use in estimating the paramagnetic susceptibility. Ferromagnetic materials begin to behave paramagnetically above their Curie temperature. In Problem 3.3 we show how to estimate the Curie temperature of a ferrimagnetic lattice by treating it as two independent ferromagnetic sublattices.

To estimate the exchange field within a magnetic material, we develop an understanding of the crystal structure for face- and body-centered cubic crystals. This allows us to introduce the concept of magnetocrystalline anisotropy and the anisotropy tensor. These concepts are explored in Problems 3.4-3.7. The Landau–Lifshitz and the Landau–Lifshitz–Gilbert equations of motion describe the response of the magnetization to an effective magnetic field. Problem 3.8 explores the relationship between these two equations, while Problems 3.9–3.10 help us relate the damped small-signal magnetic oscillations to an experimentally measured loss parameter.

© The Author(s), under exclusive license to Springer Nature Switzerland AG 2021
D. D. Stancil and A. Prabhakar, *Spin Waves*,
https://doi.org/10.1007/978-3-030-68582-9_3

Problem 3.11 introduces the concept of magnetic switching of a single-domain magnetic particle via the Stoner–Wohlfarth asteroid.

3.1   Types of Magnetic Materials

Magnetic materials are classified based on their response to the external field into

- **diamagnetic materials** that comprise of atoms with filled orbital shells, and consequently have no net magnetic moment. Such materials oppose the applied field and exhibit a negative susceptibility.
- **paramagnetic materials** that have unpaired electrons in partially filled orbitals but do not exhibit a net magnetic moment due to thermal vibrations that randomize the directions of individual magnetic moments. An applied external field causes a partial alignment of the moments and the material exhibits an induced net magnetization.
- **ferromagnetic, antiferromagnetic, and ferrimagnetic materials** that have strong exchange fields between neighboring magnetic moments, and become either parallel (ferromagnetic) or antiparallel (antiferromagnetic, ferrimagnetic) below certain transition temperatures.

3.1.1   Diamagnetism

In diamagnetic materials, application of a magnetic field induces a magnetic moment that opposes the applied field. The origin of this phenomenon can be viewed as a microscopic manifestation of Lenz's law: when a magnetic field is applied, perturbations in the orbitals of bound electrons are induced that are equivalent to microscopic circulating currents. The direction of the induced currents is such that the resulting magnetic field opposes the applied field. Unlike the motion of conduction electrons, these orbital currents remain as long as the field is applied. All materials have diamagnetic contributions to their magnetic susceptibilities, but these contributions dominate only when the material has no magnetic ions.

An approximation to the diamagnetic susceptibility arising from spherically symmetric orbitals is given by the *Langevin diamagnetism formula*:

$$\chi_{\text{dia}} = -\frac{\mu_0 N q^2}{6 m_q} \sum_{i=1}^{n} \langle r_i^2 \rangle. \tag{3.2}$$

Here N is the density of orbitals, q is the electronic charge, m_q is the electronic mass, and $\langle r_i^2 \rangle$ is the expected value of the square of the effective radius of the orbital.

3.1.2 Paramagnetism

In a paramagnetic material, the application of an external magnetic field causes a change in the *Zeeman energy* of each magnetic moment μ as

$$\mathcal{E}_{zm} = -\boldsymbol{\mu} \cdot \mathbf{B.} = +\frac{g\mu_0\mu_B}{\hbar}\mathbf{J} \cdot \mathbf{H}, = g\mu_0\mu_B J_z H/\hbar, \tag{3.3}$$

where $\mu_B = |q|\hbar/2m_q$ is the Bohr magneton, and we have assumed that the magnetic field is along the $\hat{\mathbf{z}}$ direction. Classically, the z component of angular momentum J_z can take on any value, but quantum mechanically it can only take on discrete values. The allowed values are given by $J_z = \hbar m$, where m can only have the values $-J, -J + 1, \ldots J - 1, J$, and J is an integer or half-integer representing the total angular momentum. It follows that the Zeeman energy and the z component of the magnetic moment also take on discrete values:

$$\mu_z(m) = -g\mu_B m, \tag{3.4}$$
$$\mathcal{E}_{zm}(m) = +g\mu_0\mu_B m H. \tag{3.5}$$

At a given temperature, the probability that a state with energy $\mathcal{E}$ is occupied is proportional to the *Boltzmann factor* $\exp(-\mathcal{E}/k_B T)$ where $k_B = 1.38 \times 10^{-23}$ J/K is Boltzmann's constant and T is the absolute temperature in Kelvin. Specifically, the probability of occupying the mth state can be written as

$$P(m) = Ae^{-\mathcal{E}_{zm}(m)/k_B T}, \tag{3.6}$$

where A is a normalization factor determined by requiring that

$$\sum_{m=-J}^{J} P(m) = A \sum_{m=-J}^{J} e^{-\mathcal{E}_{zm}(m)/k_B T} = 1. \tag{3.7}$$

The *thermal average* of the magnetic moment $\langle\langle \mu_z \rangle\rangle$ is obtained by multiplying the moment of the mth state by the probability that the electron is in the mth state, and summing over the allowed values of m:

$$\langle\langle \mu_z \rangle\rangle = \frac{\displaystyle\sum_{m=-J}^{J} \mu_z(m)e^{-\mathcal{E}_{zm}(m)/k_B T}}{\displaystyle\sum_{m=-J}^{J} e^{-\mathcal{E}_{zm}(m)/k_B T}} = -\frac{g\mu_B \displaystyle\sum_{m=-J}^{J} me^{-g\mu_0\mu_B m H/k_B T}}{\displaystyle\sum_{m=-J}^{J} e^{-g\mu_0\mu_B m H/k_B T}}. \tag{3.8}$$

We can obtain the thermal average of the magnetization, or dipole moment per unit volume, by multiplying the average magnetic moment by the volume density of the moments N:

$$\langle\langle M\rangle\rangle = -g\mu_B N \frac{\sum_{m=-J}^{J} me^{-mx}}{\sum_{m=-J}^{J} e^{-mx}} = g\mu_B N \frac{d}{dx}\left[\ln\left(\sum_{m=-J}^{J} e^{-mx}\right)\right], \qquad (3.9)$$

where $x = g\mu_0\mu_B H/k_B T$. The summation is a finite geometric series

$$\sum_{m=-J}^{J} e^{-mx} = \frac{e^{-(J+1/2)x} - e^{(J+1/2)x}}{e^{-x/2} - e^{x/2}} = \frac{\sinh[(J+1/2)x]}{\sinh(x/2)}. \qquad (3.10)$$

Substituting this into (3.9) and performing the indicated differentiation allows us to write

$$\langle\langle M\rangle\rangle = M_{max} B_J(\mu_0\mu_{max} H/k_B T), \qquad (3.11)$$

where $B_J(y)$ is the *Brillouin function*

$$B_J(y) = \frac{2J+1}{2J}\coth\left(\frac{2J+1}{2J}y\right) - \frac{1}{2J}\coth\left(\frac{y}{2J}\right), \qquad (3.12)$$

and

$$\mu_{max} = g\mu_B J, \qquad (3.13)$$

$$M_{max} = g\mu_B N J. \qquad (3.14)$$

For small arguments, $B_J(y)$ can be expanded in a Maclaurin series:

$$B_J(y) \approx \frac{J+1}{3J}y - \frac{[(J+1)^2 + J^2](J+1)}{90J^3}y^3 + \dots \qquad (3.15)$$

It follows that for small H/T, the average magnetization induced by the applied field H is

$$\langle\langle M\rangle\rangle = \frac{\mu_0 N g^2 J(J+1)\mu_B^2}{3k_B T} H. \qquad (3.16)$$

Since $\langle\langle M\rangle\rangle$ and H are linearly related in this approximation, we can define a paramagnetic susceptibility as

$$\chi_{para} = \frac{\langle\langle M\rangle\rangle}{H} = \frac{\mu_0 N g^2 J(J+1)\mu_B^2}{3k_B T} = \frac{C}{T}. \qquad (3.17a)$$

This form of the susceptibility with an inverse temperature dependence is called the *Curie Law*. Sometimes (3.17a) is written in terms of an *effective magneton number* $p = g[J(J+1)]^{1/2}$:

$$\chi_{para} = \frac{\mu_0 N p^2 \mu_B^2}{3k_B T}. \qquad (3.17b)$$

3.1.3 Ferromagnetism and Antiferromagnetism

In the *Weiss molecular field theory*, the total field experienced by a local magnetic moment is given by the applied field plus a *molecular field* $\mathbf{H}_m$:

$$\mathbf{H}_{tot} = \mathbf{H} + \mathbf{H}_m, \tag{3.18}$$

where the molecular field $\mathbf{H}_m$ is taken to be proportional to the local thermally-averaged magnetization:

$$\mathbf{H}_m = \lambda \mathbf{M}. \tag{3.19}$$

If $\lambda > 0$, then the lowest energy configuration has all atomic moments parallel (ferromagnetism), while if $\lambda < 0$ the lowest energy has nearest-neighbor atomic moments antiparallel (antiferromagnetism). Below a critical ordering temperature, ferromagnetic materials exhibit a net macroscopic magnetic moment, while antiferromagnets do not, owing to the cancelation of neighboring moments. The critical temperature for ferromagnets is called the *Curie temperature*, while the critical temperature for antiferromagnets is called the *Néel temperature*.

3.1.4 Ferrimagnetism

In some materials there are two sets of magnetic ions, each with a different environment in the crystal and a different magnetic moment. We refer to the two configurations as *sublattices*. Within each sublattice the interactions are ferromagnetic (all moments in the sublattice point in the same direction), whereas the interaction between the sublattices is antiferromagnetic (i.e., the two sets of moments point in opposite directions in the ground state). Such materials are called *ferrimagnets*. Because the two sublattices have different atomic moments, their contributions to the magnetization do not cancel, and ferrimagnets exhibit a net macroscopic magnetization below the Curie temperature, similar to ferromagnets. Following the *Néel theory of ferrimagnetism*, the analysis of ferrimagnets can be facilitated by considering a contribution to the molecular field from each sublattice–for example an "up" sublattice labeled a and a "down" sublattice labeled b. In this case, moments on each sublattice experience molecular fields of the form

$$\mathbf{H}_a = \lambda_{aa}\mathbf{M}_a - \lambda_{ab}\mathbf{M}_b,$$
$$\mathbf{H}_b = -\lambda_{ba}\mathbf{M}_a + \lambda_{bb}\mathbf{M}_b. \tag{3.20}$$

Here we have chosen the signs consistent with ferromagnetic coupling within each sublattice, and antiferromagnetic coupling between the sublattices.

Problem 3.1 Estimate the diamagnetic susceptibility for the two $1s$ electrons in helium using the *Langevin diamagnetism formula*:

$$\chi_{\text{dia}} = -\frac{\mu_0 N q^2}{6 m_q} \sum_{i=1}^{n} \langle r_i^2 \rangle. \tag{3.21}$$

Use the Bohr radius $r_0 = 4\pi\varepsilon_0 \hbar^2 / m_q q^2$ and $N = 2.7 \times 10^{27}$ m^3 at standard temperature and pressure (S.T.P.). Compare your result with the experimental value -1.1×10^{-7} (from Kubo, et al. [1, p. 439] converted to SI units).

Solution 3.1

The diamagnetic susceptibility given by the Langevin formula

$$\chi_{\text{dia}} = -\frac{\mu_0 N q^2}{6 m_q} \sum_{i=1}^{n} \langle r_i^2 \rangle$$

requires that we find the average orbital radius of the two $1 s$ electrons. We estimate this using the Bohr radius:

$$r_0 = \frac{4\pi\varepsilon_0 \hbar^2}{m_q q^2} = \frac{4\pi \times 8.854 \times (10^{-12}) \left[1.055 \times (10^{-34}) \right]^2}{9.11 \times (10^{-31}) \left[1.6 \times 10^{-19} \right]^2}$$

$$= 5.31 \times 10^{-11} \text{ m}.$$

At standard temperature and pressure (S.T.P.), we have $N = 2.7 \times (10^{27})$ m^{-3} and $n = 2$. Therefore,

$$\chi_{\text{dia}} = -\frac{4\pi \times 10^{-7} \times 2.7 \times 10^{27} \times \left[1.6 \times (10^{-19}) \right]^2}{6 \times (9.11) \times 10^{-31}} \times 2 \times \left[5.31 \times (10)^{-11} \right]^2$$

$$= -0.896 \times (10^{-7}).$$

$$\boxed{\chi_{\text{dia}} = -0.896 \times 10^{-7},}$$

which is in reasonable agreement with the value of $-1.1 \times (10^{-7})$ from Kubo and Nagamiya [1, p. 439].

Problem 3.2 Show that for the special case of $J=1/2$, the Brillouin function

$$B_J(y) = \frac{2J+1}{2J} \coth\left(\frac{2J+1}{2J} y \right) - \frac{1}{2J} \coth\left(\frac{y}{2J} \right) \tag{3.22}$$

can be reduced to

$$B_{1/2}(y) = \tanh y. \tag{3.23}$$

Expand this result for small y and show that the paramagnetic susceptibility for $J = 1/2$ is given by

$$\chi_{\text{para}} = \frac{\mu_0 N \mu_B^2}{k_B T}. \tag{3.24}$$

Solution 3.2

For $J = 1/2$,

$$B_J(y) = \frac{2J+1}{2J} \coth\left(\frac{2J+1}{2J} y\right) - \frac{1}{2J} \coth\left(\frac{y}{2J}\right)$$

becomes

$$B_{1/2}(y) = 2\coth(2y) - \coth y.$$

We know that

$$\coth 2y = \frac{\coth^2 y + 1}{2\coth y}.$$

Therefore,

$$B_{1/2}(y) = \frac{\coth^2 y + 1}{\coth y} - \coth y = \frac{1}{\coth y}$$

or,

$$\boxed{B_{1/2}(y) = \tanh y.}$$

For small y, $\tanh y \approx y$.

The thermal average of the magnetization is

$$\langle\langle M \rangle\rangle = g\mu_B N J B_J(g\mu_B J \mu_0 H / k_B T).$$

For $J = 1/2$ and $g = 2$, we have $gJ = 1$. Using these values along with the approximation $B_{1/2}(y) \approx y$ gives

$$\langle\langle M \rangle\rangle \approx \mu_B N \left[\mu_B \mu_0 H / k_B T\right]$$

to yield

$$\boxed{\chi_{\text{para}} = \frac{\langle\langle M \rangle\rangle}{H} = \frac{\mu_0 N \mu_B^2}{k_B T}.}$$

Problem 3.3 Show that if λ_{aa} and λ_{bb} are not neglected, the Curie temperature of a two-sublattice ferrimagnet is given by

$$T_c = \frac{1}{2}(\lambda_{aa}C_a + \lambda_{bb}C_b) + \frac{1}{2}\left[(\lambda_{aa}C_a - \lambda_{bb}C_b)^2 + 4C_aC_b\lambda_{ab}^2\right]^{1/2}. \quad (3.25)$$

Solution 3.3

For a ferrimagnet with two coupled sublattices, the sublattice magnetizations, as described in [2, Sec. 3.4]), are given by

$$M_a = M_{a,\max}B_{J_a}\left[\frac{\mu_{a,\max}\mu_0}{k_BT}(H + \lambda_{aa}M_a - \lambda_{ab}M_b)\right],$$

$$M_b = M_{b,\max}B_{J_b}\left[\frac{\mu_{b,\max}\mu_0}{k_BT}(H - \lambda_{ba}M_a + \lambda_{bb}M_b)\right].$$

Expanding the Brillouin functions to lowest order using (3.15) gives

$$M_a = \frac{C_a}{T}(H + \lambda_{aa}M_a - \lambda_{ab}M_b),$$

$$M_b = \frac{C_b}{T}(H - \lambda_{ba}M_a + \lambda_{bb}M_b),$$

where $C_{a,b} = \frac{\mu_0 N_{a,b}p_{a,b}^2\mu_B^2}{2k_B}$ and $p_{a,b} = g_{a,b}\left[J_{a,b}(J_{a,b}+1)\right]^{\frac{1}{2}}$ is the effective magneton number. The linearized equations can be written in matrix form as

$$\begin{bmatrix} T - \lambda_{aa}C_a & \lambda_{ab}C_a \\ \lambda_{ba}C_b & T - \lambda_{bb}C_b \end{bmatrix}\begin{bmatrix} M_a \\ M_b \end{bmatrix} = \begin{bmatrix} C_a \\ C_b \end{bmatrix}H.$$

Solving for $[M_a\ M_b]^T$ gives

$$\begin{bmatrix} M_a \\ M_b \end{bmatrix} = \frac{H}{(T - \lambda_{aa}C_a)(T - \lambda_{bb}C_b) - \lambda_{ab}^2C_aC_b}\begin{bmatrix} T - \lambda_{bb}C_b & -\lambda_{ab}C_a \\ -\lambda_{ba}C_b & T - \lambda_{aa}C_a \end{bmatrix}\begin{bmatrix} C_a \\ C_b \end{bmatrix}.$$

Here we have assumed that the symmetry of the interaction is such that $\lambda_{ab} = \lambda_{ba}$. The singularity in the denominator marks the onset of ferrimagnetic order and is known as the *Curie temperature*.

$$(T_c - \lambda_{aa}C_a)(T_c - \lambda_{bb}C_b) = \lambda_{ab}^2C_aC_b$$

$$\therefore\ T_c^2 - T_c(\lambda_{aa}C_a + \lambda_{bb}C_b) + (\lambda_{aa}\lambda_{bb} - \lambda_{ab}^2)C_aC_b = 0$$

Solving the quadratic equation yields two solutions, but we choose the one with the +ve sign to ensure $T_C > 0$.

$$T_c = \frac{1}{2}\left[(\lambda_{aa}C_a + \lambda_{bb}C_b) + \sqrt{(\lambda_{aa}C_a + \lambda_{bb}C_b)^2 - 4(\lambda_{aa}\lambda_{bb} - \lambda_{ab}^2)C_aC_b}\right]$$

$$= \frac{1}{2}\left[(\lambda_{aa}C_a + \lambda_{bb}C_b) + \sqrt{(\lambda_{aa}C_a)^2 + 2\lambda_{aa}\lambda_{bb}C_aC_b + (\lambda_{bb}C_b)^2 - 4\lambda_{aa}\lambda_{bb}C_aC_b + 4\lambda_{ab}^2C_aC_b}\right]$$

$$= \frac{1}{2}\left[(\lambda_{aa}C_a + \lambda_{bb}C_b) + \sqrt{(\lambda_{aa}C_a)^2 - 2\lambda_{aa}\lambda_{bb}C_aC_b + (\lambda_{bb}C_b)^2 + 4\lambda_{ab}^2C_aC_b}\right]$$

One last algebraic simplification yields

$$\boxed{T_c = \frac{1}{2}\left[(\lambda_{aa}C_a + \lambda_{bb}C_b) + \sqrt{(\lambda_{aa}C_a - \lambda_{bb}C_b)^2 + 4\lambda_{ab}^2C_aC_b}\right].}$$

3.2 Exchange Fields

To understand the origins of the molecular field theories that describe ferromagnetism and ferrimagnetism, we must return to the exchange interaction introduced in Chap. 2.

For simplicity, we consider only nearest-neighbor interactions. The total exchange energy of the ith spin interacting with its neighbors is

$$\mathcal{E}_i^{\text{ex}} = -2\sum_{j=n.n.} \mathcal{J}_{ij}\mathbf{S}_i \cdot \mathbf{S}_j = -2\mathbf{S}_i \cdot \sum_j{}' \mathcal{J}_{ij}\mathbf{S}_j. \tag{3.26}$$

Here the prime on the summation is a short-hand to remind us that the sum is only over nearest neighbors. Recognizing that the magnetic moment and spin of a particle are related by $\boldsymbol{\mu} = -g\mu_B\mathbf{S}$, we can alternatively express the exchange energy as

$$\mathcal{E}_i^{(\text{ex})} = \frac{2\boldsymbol{\mu}_i}{g\mu_B} \cdot \sum_j{}' \mathcal{J}_{ij}\mathbf{S}_j. \tag{3.27}$$

This has the same form as the Zeeman energy (3.3), provided we introduce an effective magnetic field[1] called the *exchange field*:

$$\mathbf{H}_{\text{ex}} = -\frac{2}{g\mu_0\mu_B} \sum_j{}' \mathcal{J}_{ij}\mathbf{S}_j. \tag{3.28}$$

---

[1] In general, the ith component of the effective magnetic field is defined as $H_{\text{eff},i} = -(\delta\mathcal{E}/\delta\mu_i)/\mu_0$, where $\delta\mathcal{E}/\delta\mu_i$ is the variational derivative defined by (7.2).

3.3   Crystal Structures

Crystalline materials have an underlying periodic geometry in the arrangement of atoms. The fundamental building block is a unit cell, that is then repeated in all 3 dimensions. Of interest to us are the *body-centered cubic* and the *face-centered cubic* cells, commonly referred to as bcc and fcc respectively, and shown in Fig. 3.1.

The unit cell in a bcc lattice comprises of an atom at each of the 8 vertices of a cube, with one more at the center of the cell. Similarly, the unit cell in an fcc structure has 8 atoms at the vertices of a cube, and one at the center of each face of the cube. The distance between an atom and its nearest neighbor is thus determined by the length of the edge of the cube, and the underlying atomic arrangement within a unit cell.

3.3.1   Miller Indices

The periodic arrangement of atoms in the crystal forms a lattice. We describe certain directions and planes within the lattices using a set of three integers, or the *Miller indices*. For a cubic unit cell with an edge of length a, the $[hkl]$ indices describe the orientation of a family of planes with spacing d_{hkl} given by

$$\frac{1}{d_{hkl}^2} = \frac{h^2 + k^2 + l^2}{a^2} \tag{3.29}$$

Thus, [100] refers to the planes oriented with the normal to the plane along x, and a spacing of a between them. Similarly, the [110] planes have a spacing of $a/\sqrt{2}$ between them.

Problem 3.4 Verify the equation

$$\sum_{j}' S(\mathbf{r}_i + {}_j) = ZS(\mathbf{r}_i) + \frac{ZR_n^2}{6} \nabla^2 S(\mathbf{r}_i), \tag{3.30}$$

for face-centered and body-centered cubic lattices.

Fig. 3.1 Unit cells for the face-centered and body-centered cubic crystals

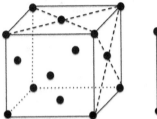 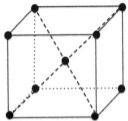

Solution 3.4

We need to verify that

$$\sum_j \overline{S}\left(\overline{r}_i + \overline{\delta}_j\right) = Z\overline{S}\left(\overline{r}_i\right) + \frac{ZR_n^2}{6}\nabla^2 \overline{S}\left(\overline{r}_i\right)$$

where Z is the number of nearest neighbors and R_n is the distance to the nearest neighbor.

fcc

An atom at the corner of the unit cell has 12 nearest neighbors, one at the center of each face of the cube [3]. Thus, $Z = 12$, $R_n = \frac{a}{\sqrt{2}}$ (a = cube edge), and the nearest neighbor locations are

$$\frac{a}{2}\left(\pm\hat{x} \pm \hat{y}\right), \frac{a}{2}\left(\pm\hat{y} \pm \hat{z}\right), \frac{a}{2}\left(\pm\hat{x} \pm \hat{z}\right).$$

Using a *Taylor series* expansion we can write:

$$\overline{S}\left(\overline{r} + \overline{\delta}\right) = \overline{S}\left(\overline{r}\right) + \left(\overline{\delta}\cdot\nabla\right)\overline{S}\left(\overline{r}\right) + \frac{\left(\overline{\delta}\cdot\nabla\right)^2}{2!}\overline{S}\left(\overline{r}\right) + ...$$

Thus

$$\overline{S}\left(\overline{r}_i + \alpha\hat{x} + \beta\hat{y}\right) \approx \overline{S}\left(\overline{r}_i\right) + \alpha\frac{\partial\overline{S}\left(\overline{r}_i\right)}{\partial x} + \beta\frac{\partial\overline{S}\left(\overline{r}_i\right)}{\partial y}$$
$$+ \frac{\alpha^2}{2}\frac{\partial^2\overline{S}}{\partial x^2} + \frac{\beta^2}{2}\frac{\partial^2\overline{S}}{\partial y^2} + \alpha\beta\frac{\partial^2\overline{S}}{\partial x\partial y} + ...$$

It follows that

$$\overline{S}\left(\overline{r}_i + \alpha\hat{x} + \beta\hat{y}\right) + \overline{S}\left(\overline{r}_i - \alpha\hat{x} - \beta\hat{y}\right) \approx 2\overline{S}\left(\overline{r}_i\right) + \alpha^2\frac{\partial^2\overline{S}}{\partial x^2} + \beta^2\frac{\partial^2\overline{S}}{\partial y^2}$$
$$+ 2\alpha\beta\frac{\partial^2\overline{S}}{\partial x\partial y} \tag{3.31}$$

and

$$\overline{S}\left(\overline{r}_i + \alpha\hat{x} - \beta\hat{y}\right) + \overline{S}\left(\overline{r}_i - \alpha\hat{x} + \beta\hat{y}\right) \approx 2\overline{S}\left(\overline{r}_i\right) + \alpha^2\frac{\partial^2\overline{S}}{\partial x^2} + \beta^2\frac{\partial^2\overline{S}}{\partial y^2}$$
$$- 2\alpha\beta\frac{\partial^2\overline{S}}{\partial x\partial y}, \tag{3.32}$$

to second order in the expansion. Adding (3.31) and (3.32) gives

$$\Sigma\overline{S}\left(\overline{r}_i \pm \alpha\hat{x} \pm \beta\hat{y}\right) \approx 4\overline{S}\left(\overline{r}_i\right) + 2\alpha^2\frac{\partial^2\overline{S}}{\partial x^2} + 2\beta^2\frac{\partial^2\overline{S}}{\partial y^2}.$$

Substituting this result into the sum over nearest neighbors gives

$$\sum_j \overline{S}(\overline{r_i} + \overline{\delta_j}) = 4\overline{S}(\overline{r_i}) + \frac{a^2}{2}\frac{\partial^2 \overline{S}}{\partial x^2} + \frac{a^2}{2}\frac{\partial^2 \overline{S}}{\partial y^2}$$

$$+ 4\overline{S}(\overline{r_i}) + \frac{a^2}{2}\frac{\partial^2 \overline{S}}{\partial y^2} + \frac{a^2}{2}\frac{\partial^2 \overline{S}}{\partial z^2}$$

$$+ 4\overline{S}(\overline{r_i}) + \frac{a^2}{2}\frac{\partial^2 \overline{S}}{\partial x^2} + \frac{a^2}{2}\frac{\partial^2 \overline{S}}{\partial z^2}$$

$$\sum_j \overline{S}(\overline{r_i} + \overline{\delta_j}) = \underbrace{12}_{Z}\, \overline{S}(\overline{r_i}) + \underbrace{a^2}_{\frac{ZR_n^2}{6} = \frac{12}{6}\frac{a^2}{2} = a^2} \left(\frac{\partial^2 \overline{S}}{\partial x^2} + \frac{\partial^2 \overline{S}}{\partial y^2} + \frac{\partial^2 \overline{S}}{\partial z^2} \right)$$

$$\boxed{\text{Form is correct}}$$

bcc

For an atom at the corner of a unit cell the nearest neighbor is at the center of the cube, and there are 8 such neighbors for each atom. Thus, $Z = 8$ and $R_n = a\frac{\sqrt{3}}{2}$ with the nearest neighbors located at $\frac{a}{2}(\pm\hat{x} \pm \hat{y} \pm \hat{z})$.

$$\sum_j \overline{S}(\overline{r_i} + \overline{\delta_j}) = \overline{S}\left(\overline{r_i} + \frac{a}{2}\hat{x} + \frac{a}{2}\hat{y} + \frac{a}{2}\hat{z}\right) + \overline{S}\left(\overline{r_i} - \frac{a}{2}\hat{x} - \frac{a}{2}\hat{y} - \frac{a}{2}\hat{z}\right)$$

$$+ \overline{S}\left(\overline{r_i} + \frac{a}{2}\hat{x} + \frac{a}{2}\hat{y} - \frac{a}{2}\hat{z}\right) + \overline{S}\left(\overline{r_i} - \frac{a}{2}\hat{x} - \frac{a}{2}\hat{y} + \frac{a}{2}\hat{z}\right)$$

$$+ \overline{S}\left(\overline{r_i} + \frac{a}{2}\hat{x} - \frac{a}{2}\hat{y} + \frac{a}{2}\hat{z}\right) + \overline{S}\left(\overline{r_i} - \frac{a}{2}\hat{x} + \frac{a}{2}\hat{y} - \frac{a}{2}\hat{z}\right)$$

$$+ \overline{S}\left(\overline{r_i} - \frac{a}{2}\hat{x} + \frac{a}{2}\hat{y} + \frac{a}{2}\hat{z}\right) + \overline{S}\left(\overline{r_i} + \frac{a}{2}\hat{x} - \frac{a}{2}\hat{y} - \frac{a}{2}\hat{z}\right)$$

$$= 2\overline{S}(\overline{r_i}) + \frac{a^2}{4}\frac{\partial^2 \overline{S}}{\partial x^2} + \frac{a^2}{4}\frac{\partial^2 \overline{S}}{\partial y^2} + \frac{a^2}{4}\frac{\partial^2 \overline{S}}{\partial z^2} + \frac{a^2}{2}\frac{\partial^2 \overline{S}}{\partial x \partial y} + \frac{a^2}{2}\frac{\partial^2 \overline{S}}{\partial x \partial z} + \frac{a^2}{2}\frac{\partial^2 \overline{S}}{\partial y \partial z}$$

$$+ 2\overline{S}(\overline{r_i}) + \frac{a^2}{4}\frac{\partial^2 \overline{S}}{\partial x^2} + \frac{a^2}{4}\frac{\partial^2 \overline{S}}{\partial y^2} + \frac{a^2}{4}\frac{\partial^2 \overline{S}}{\partial z^2} + \frac{a^2}{2}\frac{\partial^2 \overline{S}}{\partial x \partial y} - \frac{a^2}{2}\frac{\partial^2 \overline{S}}{\partial x \partial z} - \frac{a^2}{2}\frac{\partial^2 \overline{S}}{\partial y \partial z}$$

$$+ 2\overline{S}(\overline{r_i}) + \frac{a^2}{4}\frac{\partial^2 \overline{S}}{\partial x^2} + \frac{a^2}{4}\frac{\partial^2 \overline{S}}{\partial y^2} + \frac{a^2}{4}\frac{\partial^2 \overline{S}}{\partial z^2} - \frac{a^2}{2}\frac{\partial^2 \overline{S}}{\partial x \partial y} + \frac{a^2}{2}\frac{\partial^2 \overline{S}}{\partial x \partial z} - \frac{a^2}{2}\frac{\partial^2 \overline{S}}{\partial y \partial z}$$

$$+ 2\overline{S}(\overline{r_i}) + \frac{a^2}{4}\frac{\partial^2 \overline{S}}{\partial x^2} + \frac{a^2}{4}\frac{\partial^2 \overline{S}}{\partial y^2} + \frac{a^2}{4}\frac{\partial^2 \overline{S}}{\partial z^2} - \frac{a^2}{2}\frac{\partial^2 \overline{S}}{\partial x \partial y} - \frac{a^2}{2}\frac{\partial^2 \overline{S}}{\partial x \partial z} + \frac{a^2}{2}\frac{\partial^2 \overline{S}}{\partial y \partial z}$$

These all add $\qquad\qquad\qquad\qquad$ These all cancel

$$\therefore \sum_j \overline{S}(\overline{r_i} + \overline{\delta_j}) = \underbrace{8}_{Z}\, \overline{S}(\overline{r_i}) + \underbrace{a^2}_{\frac{ZR_n^2}{6} = \frac{8}{6}a^2\frac{3}{4} = a^2} \left(\frac{\partial^2 \overline{S}}{\partial x^2} + \frac{\partial^2 \overline{S}}{\partial y^2} + \frac{\partial^2 \overline{S}}{\partial z^2} \right)$$

$$\boxed{\text{Form is correct}}$$

3.4   Magnetocrystalline Anisotropy

In general, the orientation of an atomic orbital that is not spherically symmetric will be affected by interactions with neighboring ions in the crystal. In such cases, the interaction between the spin and orbital magnetic moments (spin–orbit coupling) causes the energy of the net moment of the ion to depend on its orientation with respect to the crystal axes. This orientation dependence is referred to as the *magnetocrystalline anisotropy* energy, $\mathcal{E}_a$.

In the case of *uniaxial anisotropy*, there is a particular direction in the crystal along which the energy is either higher (a "hard" axis) or lower (an "easy" axis). Because of the crystal symmetry, there is no difference between an orientation that is "up" or "down" along a symmetry axis. As a result, the anisotropy energy density can be expressed in terms of even powers of the magnetization component along the symmetry axis:

$$W_{\mathrm{au}} = K_{u1}(1 - M_3^2/M_S^2) + K_{u2}(1 - M_3^2/M_S^2)^2 + \dots . \tag{3.33}$$

where M_3 is the component of $\mathbf{M}$ along the symmetry axis, and M_S is the saturation magnetization.

For *cubic anisotropy*, we again require that reversal of $\mathbf{M}$ must not affect the energy, so that the energy density must be expressed in terms of even powers of the magnetization components. Further, considerations of crystal symmetry yield the following additional condition: interchanging any two axes should not change W_a.

A function that satisfies these two conditions to sixth order is

$$W_{\mathrm{ac}} = \frac{K_{c1}}{M_{S^4}}[M_1^2 M_2^2 + M_2^2 M_3^2 + M_3^2 M_1^2] + \frac{K_{c2}}{M_{S^6}} M_1^2 M_2^2 M_3^2 + \dots . \tag{3.34}$$

Since the energy of a magnetic moment also depends on orientation when a magnetic field is applied, for small perturbations we can model the anisotropy energy as resulting from an *effective anisotropy field*, $\mathbf{H}_k$, defined as

$$\mathbf{H}_k = -\frac{1}{\mu_0}\frac{\partial W_a}{\partial \mathbf{M}}. \tag{3.35}$$

As an example, to lowest order, the effective cubic anisotropy field is

$$
\begin{aligned}
H_{kc} &= -\frac{K_{c1}}{\mu_0 M_S^4}\frac{\partial}{\partial \mathbf{M}}[M_1^2 M_2^2 + M_2^2 M_3^2 + M_3^2 M_1^2] \\
&= -\frac{2K_{c1}}{\mu_0 M_S^4}\begin{bmatrix} M_1(M_2^2 + M_3^2) \\ M_2(M_3^2 + M_1^2) \\ M_3(M_1^2 + M_2^2) \end{bmatrix} .
\end{aligned}
\tag{3.36}
$$

For small perturbations from equilibrium, we can express the magnetization as a static component plus a small time-varying component:

$$\mathbf{M} = \mathbf{M}_0 + \mathbf{m}(t). \tag{3.37}$$

Substituting (3.37) into (3.36) and keeping only linear terms in $\mathbf{m}$ gives[4]

$$H_{kc} = -\frac{2K_{c1}}{\mu_0 M_S^4} \begin{bmatrix} M_{01}(M_{02}^2 + M_{03}^2) \\ M_{02}(M_{03}^2 + M_{01}^2) \\ M_{03}(M_{01}^2 + M_{02}^2) \end{bmatrix} + \overline{\mathbf{N}}^a \cdot \mathbf{m}, \qquad (3.38)$$

where the off-diagonal elements of the anisotropy tensor $\overline{\mathbf{N}}^a$ are

$$N_{ij}^a = -\frac{4K_{c1}}{\mu_0 M_S^4} M_{0i} M_{0j}; \qquad i \neq j, \qquad (3.39a)$$

and the diagonal elements are

$$N_{ii}^a = -\frac{4K_{c1}}{\mu_0 M_S^4} [M_S^2 - M_{0i}^2]. \qquad (3.39b)$$

3.4.1   Coordinate Transformations

Sometimes it is desirable to work in a different coordinate system than the crystal coordinate system, e.g., if a different coordinate system simplifies a calculation or better matches a laboratory experiment. In such cases we would like to be able to take the fields and tensors in one system and transform them into another. For example, if $\hat{\mathbf{e}}_k$ is the kth unit basis vector in the original system, we can generate basis unit vectors in a different system $\hat{\mathbf{e}}_j'$ using

$$\hat{\mathbf{e}}_j' = \sum_k T_{jk} \hat{\mathbf{e}}_k \qquad (3.40)$$

where the jk th element of the transformation matrix $\overline{\mathbf{T}}$ is given by

$$T_{jk} = \hat{\mathbf{e}}_j' \cdot \hat{\mathbf{e}}_k. \qquad (3.41)$$

Using the transformation matrix $\overline{\mathbf{T}}$, an arbitrary vector $\mathbf{a}$ in the original system can be transformed into the primed system:

$$\mathbf{a}' = \overline{\mathbf{T}} \cdot \mathbf{a}. \qquad (3.42)$$

Similarly, an arbitrary tensor in the original system can be transformed into the primed system using

$$\overline{\mathbf{N}}' = \overline{\mathbf{T}} \cdot \overline{\mathbf{N}} \cdot \overline{\mathbf{T}}^T. \qquad (3.43)$$

Problem 3.5 Verify that the second-order cubic anisotropy field along the [111] direction is given by

$$\mathbf{H}_{0k}^{(2)'} = -\frac{2K_{c2}}{9\mu_0 M_S}\hat{e}_3' \tag{3.44}$$

where $\hat{e}_3'$ is a unit vector along the [111] direction.

Solution 3.5

To verify that the second-order cubic anisotropy field along the [111] direction is given by (3.44), we start with the energy term

$$W_{ac}^{(2)} = \frac{K_{c2}}{M_S^6} M_1^2 M_2^2 M_3^2$$

$$\therefore \overline{H}_{kc}^{(2)} = -\frac{1}{\mu_0}\frac{\partial}{\partial \overline{M}} W_{ac}^{(2)} = -\frac{K_{c2}}{\mu_0 M_S^6}\frac{\partial}{\partial \overline{M}}(M_1^2 M_2^2 M_3^2)$$

$$= \frac{-K_{c2}}{\mu_0 M_S^6}\left[\hat{e}_1 2M_1 M_2^2 M_3^2 + \hat{e}_2 2M_1^2 M_2 M_3^2 + \hat{e}_3 2M_1^2 M_2^2 M_3\right].$$

For $\overline{M} \parallel [111]$, $M_1 = M_2 = M_3 = \dfrac{M_S}{\sqrt{3}}$.

$$\therefore \overline{H}_{kc}^{(2)} = -\frac{K_{c2}}{\mu_0 M_S}\frac{2}{3^{\frac{5}{2}}}\left[\hat{e}_1 + \hat{e}_2 + \hat{e}_3\right]$$

$$= \underbrace{\frac{-2K_{c2}}{9\mu_0 M_S}}_{\text{(Magnitude)}} \underbrace{\frac{1}{\sqrt{3}}\left[\hat{e}_1 + \hat{e}_2 + \hat{e}_3\right]}_{\text{(Unit vector }=\hat{e}_3'\text{ in new coordinate system)}}$$

Hence,

$$\boxed{\overline{H}_{kc}^{(2)} = \frac{-2K_{c2}}{9\mu_0 M_S}\hat{e}_3'}$$

Problem 3.6 A cubic crystal is magnetized along the $[\bar{1}10]$ direction.

(a) Show that the first-order static anisotropy field is also along the $[\bar{1}10]$ direction with magnitude $-K_{c1}/\mu_0 M_S$.
(b) Show that the first-order anisotropy tensor in the crystal coordinate system is given by

$$\overline{N}^a = \frac{K_{c1}}{\mu_0 M_S^2}\begin{bmatrix} -1 & 2 & 0 \\ 2 & -1 & 0 \\ 0 & 0 & -2 \end{bmatrix}. \tag{3.45}$$

(c) Transform the anisotropy tensor of part (b) into a coordinate system where $\hat{z} \parallel [\bar{1}10]$ and $\hat{y} \parallel [111]$. Show that the result is

$$\overline{\mathbf{N}}^{\mathrm{a}'} = \frac{K_{c1}}{\mu_0 M_S^2} \begin{bmatrix} -1 & \sqrt{2} & 0 \\ -\sqrt{2} & 0 & 0 \\ 0 & 0 & -3 \end{bmatrix}. \tag{3.46}$$

Solution 3.6

A cubic crystal is magnetized along the $[\bar{1}10]$ direction.

(a) Show that the first-order static anisotropy field is also along the $[\bar{1}10]$ direction with magnitude $-K_{c1}/\mu_0 M_S$. In this case,

$$\overline{M} \parallel [\bar{1}10], \quad \overline{M} = \frac{M_S}{\sqrt{2}} [-\hat{e}_1 + \hat{e}_2].$$

It follows that

$$\overline{H}_{k0} = \frac{-2K_{c1}}{\mu_0 M_S^4} \begin{bmatrix} M_{01}(M_{02}^2 + M_{03}^2) \\ M_{02}(M_{01}^2 + M_{03}^2) \\ M_{03}(M_{01}^2 + M_{02}^2) \end{bmatrix} = \frac{-2K_{c1}}{\mu_0 M_S^4} \begin{bmatrix} \frac{-M_S^3}{2\sqrt{2}} \\ \frac{M_S^3}{2\sqrt{2}} \\ 0 \end{bmatrix}$$

$$\boxed{\overline{H}_{k0} = \underbrace{\frac{-K_{c1}}{\mu_0 M_S}}_{\text{magnitude}} \underbrace{\frac{1}{\sqrt{2}} \begin{bmatrix} -1 \\ 1 \\ 0 \end{bmatrix}}_{\text{unit vector } [\bar{1}10]}}$$

(b) The first-order anisotropy tensor in the crystal coordinate system is given by (from 3.39)

$$\overline{\overline{N}}^{\mathrm{a}} = \frac{-2K_{c1}}{\mu_0 M_S^4} \begin{bmatrix} \frac{M_S^2}{2} & -M_S^2 & 0 \\ -M_S^2 & \frac{M_S^2}{2} & 0 \\ 0 & 0 & M_S^2 \end{bmatrix}$$

which simplifies to

$$\boxed{\overline{\overline{N}}^{\mathrm{a}} = \frac{K_{c1}}{\mu_0 M_S^2} \begin{bmatrix} -1 & 2 & 0 \\ 2 & -1 & 0 \\ 0 & 0 & -2 \end{bmatrix}.}$$

(c) When we transform the anisotropy tensor of part (b) into a coordinate system where $\hat{z} \parallel [\bar{1}10]$ and $\hat{y} \parallel [111]$, we get the unit vectors

$$\hat{z} = \frac{1}{\sqrt{2}} \left(-\hat{e}_1 + \hat{e}_2 \right)$$

$$\hat{y} = \frac{1}{\sqrt{3}} \left(\hat{e}_1 + \hat{e}_2 + \hat{e}_3 \right)$$

$$\text{Thus } \hat{x} = \hat{y} \times \hat{z} = \frac{1}{\sqrt{6}} \left(-\hat{e}_1 - \hat{e}_2 + 2\hat{e}_3 \right)$$

From (3.41), the jk th element of the transformation matrix is therefore $T_{jk} = \hat{j} \cdot \hat{e}_k$, where $j = x, y, z$, and $k = 1, 2, 3$:

$$\overline{\overline{T}} = \begin{bmatrix} \frac{-1}{\sqrt{6}} & \frac{-1}{\sqrt{6}} & \frac{2}{\sqrt{6}} \\ \frac{1}{\sqrt{3}} & \frac{1}{\sqrt{3}} & \frac{1}{\sqrt{3}} \\ \frac{-1}{\sqrt{2}} & \frac{1}{\sqrt{2}} & 0 \end{bmatrix}$$

$$\overline{\overline{N}}^{a'} = \overline{\overline{T}} \cdot \overline{\overline{N}}^a \cdot \overline{\overline{T}}^T = C\overline{\overline{T}} \cdot \begin{bmatrix} -1 & 2 & 0 \\ 2 & -1 & 0 \\ 0 & 0 & -2 \end{bmatrix} \begin{bmatrix} \frac{-1}{\sqrt{6}} & \frac{1}{\sqrt{3}} & \frac{-1}{\sqrt{2}} \\ \frac{-1}{\sqrt{6}} & \frac{1}{\sqrt{3}} & \frac{1}{\sqrt{2}} \\ \frac{2}{\sqrt{6}} & \frac{1}{\sqrt{3}} & 0 \end{bmatrix}$$

$$= C \begin{bmatrix} \frac{-1}{\sqrt{6}} & \frac{-1}{\sqrt{6}} & \frac{2}{\sqrt{6}} \\ \frac{1}{\sqrt{3}} & \frac{1}{\sqrt{3}} & \frac{1}{\sqrt{3}} \\ \frac{-1}{\sqrt{2}} & \frac{1}{\sqrt{2}} & 0 \end{bmatrix} \begin{bmatrix} \frac{-1}{\sqrt{6}} & \frac{1}{\sqrt{3}} & \frac{3}{\sqrt{2}} \\ \frac{-1}{\sqrt{6}} & \frac{1}{\sqrt{3}} & \frac{-3}{\sqrt{2}} \\ \frac{-4}{\sqrt{6}} & \frac{-2}{\sqrt{3}} & 0 \end{bmatrix}$$

Finally,

$$\boxed{\overline{\overline{N}}^{a'} = \frac{K_{c1}}{\mu_0 M_S^2} \begin{bmatrix} -1 & -\sqrt{2} & 0 \\ -\sqrt{2} & 0 & 0 \\ 0 & 0 & -3 \end{bmatrix},}$$

with

$$C = \frac{K_{c1}}{\mu_0 M_S^2}.$$

Problem 3.7 Consider a ferrite magnetized in the plane perpendicular to the [111] direction.

(a) Show that the magnetization vector making the angle ψ with the $[\bar{1}10]$ direction is

$$\mathbf{M} = M_S \begin{bmatrix} -\frac{1}{\sqrt{2}} \cos\psi - \frac{1}{\sqrt{6}} \sin\psi \\ \frac{1}{\sqrt{2}} \cos\psi - \frac{1}{\sqrt{6}} \sin\psi \\ +\sqrt{\frac{2}{3}} \sin\psi \end{bmatrix} \qquad (3.47)$$

Fig. 3.2 Geometry for the
magnetization in the $x - y$
plane, perpendicular to [111]

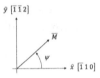

in the crystal coordinate system.

(b) Show that the first-order cubic anisotropy energy is independent of the angle ψ.

Solution 3.7

For a ferrite magnetized in the plane perpendicular to the [111] direction, show that
the magnetization vector making the angle ψ with the [$\bar{1}$10] direction in the crystal
coordinate system is

$$\overline{M} = M_S \begin{bmatrix} -\frac{1}{\sqrt{2}}\cos\psi - \frac{1}{\sqrt{6}}\sin\psi \\ \frac{1}{\sqrt{2}}\cos\psi - \frac{1}{\sqrt{6}}\sin\psi \\ +\sqrt{\frac{2}{3}}\sin\psi \end{bmatrix}$$

(a) Two orthogonal vectors perpendicular to [111] are [$\bar{1}$10] and [$\bar{1}\bar{1}$2], which we
write as

$$\hat{x} = \frac{1}{\sqrt{2}}(-\hat{e}_1 + \hat{e}_2),$$

$$\hat{y} = \frac{1}{\sqrt{6}}(-\hat{e}_1 - \hat{e}_2 + 2\hat{e}_3).$$

$\overline{M}$ can be written in terms of its component vectors in Cartesian coordinates,
and then expressed in the crystal coordinates as (see Fig. 3.2)

$$\overline{M} = M_S(\hat{x}\cos\psi + \hat{y}\sin\psi)$$
$$= M_S\left(-\hat{e}_1\frac{\cos\psi}{\sqrt{2}} + \hat{e}_2\frac{\cos\psi}{\sqrt{2}} - \hat{e}_1\frac{\sin\psi}{\sqrt{6}} - \hat{e}_2\frac{\sin\psi}{\sqrt{6}} + 2\hat{e}_3\frac{\sin\psi}{\sqrt{6}}\right)$$
$$= M_S\left[-\hat{e}_1\left(\frac{1}{\sqrt{2}}\cos\psi + \frac{1}{\sqrt{6}}\sin\psi\right) + \hat{e}_2\left(\frac{\cos\psi}{\sqrt{2}} - \frac{\sin\psi}{\sqrt{6}}\right) + \hat{e}_3\frac{2\sin\psi}{\sqrt{6}}\right]$$

which we rewrite as the column matrix

$$\overline{M} = M_S \begin{bmatrix} -\frac{1}{\sqrt{2}}\cos\psi - \frac{1}{\sqrt{6}}\sin\psi \\ \frac{1}{\sqrt{2}}\cos\psi - \frac{1}{\sqrt{6}}\sin\psi \\ \frac{2}{\sqrt{6}}\sin\psi \end{bmatrix}.$$

(b) To show that the first-order cubic anisotropy energy is independent of the angle ψ, we begin with the energy term

$$W^{(1)} = \frac{K_{c1}}{M_S^4}[M_{01}^2 M_{02}^2 + M_{01}^2 M_{03}^2 + M_{02}^2 M_{03}^2].$$

Now, we use the direction cosines to simplify each term:

$$
\begin{aligned}
\frac{M_{01}^2 M_{02}^2}{M_S^4} &= \left(\frac{\cos\psi}{\sqrt{2}} + \frac{\sin\psi}{\sqrt{6}}\right)^2 \left(\frac{\cos\psi}{\sqrt{2}} - \frac{\sin\psi}{\sqrt{6}}\right)^2 \\
&= \left(\frac{\cos^2\psi}{2} + \frac{2\cos\psi\sin\psi}{\sqrt{12}} + \frac{\sin^2\psi}{6}\right) \\
&\quad \times \left(\frac{\cos^2\psi}{2} - \frac{2\cos\psi\sin\psi}{\sqrt{12}} + \frac{\sin^2\psi}{6}\right) \\
&= \left(\frac{\cos^2\psi}{2} + \frac{\sin^2\psi}{6}\right)^2 - \frac{4\cos^2\psi\sin^2\psi}{12} \\
&= \left(\frac{1}{6} + \frac{1}{3}\cos^2\psi\right)^2 - \frac{\cos^2\psi\sin^2\psi}{3}.
\end{aligned}
\tag{3.48a}
$$

$$
\frac{M_{01}^2 M_{03}^2}{M_S^4} = \left(\frac{\cos^2\psi}{2} + \frac{2\cos\psi\sin\psi}{\sqrt{12}} + \frac{\sin^2\psi}{6}\right)\frac{2\sin^2\psi}{3}.
\tag{3.48b}
$$

$$
\frac{M_{02}^2 M_{03}^2}{M_S^4} = \left(\frac{\cos^2\psi}{2} - \frac{2\cos\psi\sin\psi}{\sqrt{12}} + \frac{\sin^2\psi}{6}\right)\frac{2\sin^2\psi}{3}.
\tag{3.48c}
$$

Thus,

$$
\begin{aligned}
\frac{1}{M_S^4}\left(M_{01}^2 M_{03}^2 + M_{02}^2 M_{03}^2\right) &= \left(\frac{\cos^2\psi}{2} + \frac{\sin^2\psi}{6}\right)\frac{4\sin^2\psi}{3} \\
&= \left(\frac{1}{6} + \frac{1}{3}\cos^2\psi\right)\frac{4}{3}\sin^2\psi.
\end{aligned}
$$

Adding in (3.48a) gives

$$
\begin{aligned}
\frac{1}{M_S^4}&(M_{01}^2 M_{02}^2 + M_{01}^2 M_{03}^2 + M_{02}^2 M_{03}^2) \\
&= \left(\frac{1}{6} + \frac{1}{3}\cos^2\psi\right)^2 - \frac{\cos^2\psi\sin^2\psi}{3} + \left(\frac{1}{6} + \frac{1}{3}\cos^2\psi\right)\frac{4}{3}\sin^2\psi \\
&= \left(\frac{1}{6} + \frac{1}{3}\cos^2\psi\right)\left(\frac{1}{2} + \sin^2\psi\right) - \frac{\cos^2\psi\sin^2\psi}{3} \\
&= \frac{1}{12} + \frac{1}{6}\cos^2\psi + \frac{1}{6}\sin^2\psi + \frac{\cos^2\psi\sin^2\psi}{3} - \frac{\cos^2\psi\sin^2\psi}{3} \\
&= \frac{1}{4}.
\end{aligned}
$$

Thus $W^{(1)} = \dfrac{K_{c1}}{4}$ is independent of ψ.

3.5   Equation of Motion for the Magnetization

Using the analogy of a spinning top from Chap. 1, we write the equation of motion for the angular momentum $\mathbf{J}$ in the presence of a magnetic field as (cf. (1.1)):

$$\frac{d\mathbf{J}}{dt} = \gamma\mu_0\mathbf{J} \times \mathbf{H}_{\text{eff}}, \tag{3.49}$$

where $\mathbf{H}_{\text{eff}}$ represents the sum of all torque-producing effective fields:

$$\mathbf{H}_{\text{eff}} = \mathbf{H} + \mathbf{H}_{\text{ex}} + \mathbf{H}_k . \tag{3.50}$$

Here $\mathbf{H}$ is the Maxwellian field, $\mathbf{H}_{\text{ex}}$ is the exchange field (3.28) and $\mathbf{H}_k$ is the anisotropy field (3.35). Noting that $\mathbf{M} = \gamma N\mathbf{J}$, we obtain the lossless form of the *Landau–Lifshitz* equation

$$\frac{d\mathbf{M}}{dt} = \gamma\mu_0\mathbf{M} \times \mathbf{H}_{\text{eff}}. \tag{3.51}$$

The lossless form of the Landau–Lifshitz equation is also commonly called the *torque equation*.

In (3.28) the exchange field is expressed in terms of the spin. To express it in terms of the magnetization approximated as a continuous function, we substitute (3.30) into (3.28) and again use the relation $\mathbf{M} = \gamma N\mathbf{S} = -g\mu_B N\mathbf{S}$ to obtain

$$\mathbf{H}_{\text{ex}} = \frac{2ZJ}{\mu_0 Ng^2\mu_B^2}\left[\mathbf{M} + \frac{R_n^2}{6}\nabla^2\mathbf{M}\right]. \tag{3.52}$$

Comparing (3.52) with (3.51) and noting that $\mathbf{M} \times \mathbf{M} \equiv 0$, we see that to understand the dynamics of $\mathbf{M}$, it is sufficient to keep only the second term in (3.52):

$$\mathbf{H}_{\text{ex}} = \lambda_{\text{ex}}\nabla^2\mathbf{M}, \tag{3.53}$$

where $\lambda_{\text{ex}} = 2ZJR_n^2/(6\mu_0 Ng^2\mu_B^2)$ for a cubic ferromagnet[2] (cf. (3.52)). The effective anisotropy field $\mathbf{H}_k$ is obtained for a given geometry and crystal structure as discussed in the previous section.

---

[2]Equation (3.53) is also applicable to a cubic ferrimagnet, but the relation between λ_{ex} and the microscopic parameters is more complicated.

3.6   Susceptibility Without Exchange or Anisotropy

As we did previously in our discussion of the effective anisotropy field, let us again express the field quantities as small time-varying perturbations about an equilibrium value as follows:

$$\mathbf{M} = \mathbf{M}_0 + \mathbf{m}(t), \tag{3.54}$$

$$\mathbf{H} = \mathbf{H}_0 + \mathbf{h}(t). \tag{3.55}$$

For simplicity, we have neglected $\mathbf{H}_{ex}$ and $\mathbf{H}_k$. Substituting these equations into the equation of motion (3.51) gives

$$\frac{d\mathbf{m}}{dt} = \gamma\mu_0[\mathbf{M}_0 \times \mathbf{H}_0 + \mathbf{M}_0 \times \mathbf{h} + \mathbf{m} \times \mathbf{H}_0 + \mathbf{m} \times \mathbf{h}]. \tag{3.56}$$

Let us assume that the applied field is strong enough to create a single-domain material with $\mathbf{M}_0$ parallel to $\mathbf{H}_0$. In this case the first term in (3.56) vanishes. Also, since we are taking the magnitudes of $\mathbf{m}$ and $\mathbf{h}$ to be small compared with $\mathbf{M}_0$ and $\mathbf{H}_0$, we can neglect the last term since it is second-order in small quantities.

Without loss of generality, we can take the equilibrium direction to be along $\hat{\mathbf{z}}$. The magnitude of the static component of the magnetization will then be $M_0 = M_z = M_S \cos\theta \approx M_S$. Further, if the time dependence is given by $\exp(-i\omega t)$, the small-signal equation of motion (also known as the "torque equation") becomes

$$-i\omega\mathbf{m} = \hat{\mathbf{z}} \times [-\omega_M\mathbf{h} + \omega_0\mathbf{m}], \tag{3.57}$$

where

$$\omega_M \equiv -\gamma\mu_0 M_S \tag{3.58}$$

and

$$\omega_0 \equiv -\gamma\mu_0 H_0. \tag{3.59}$$

The linearized torque equation (3.57) can be solved for $\mathbf{h}$ to obtain

$$\begin{bmatrix} h_x \\ h_y \end{bmatrix} = \frac{1}{\omega_M}\begin{bmatrix} \omega_0 & i\omega \\ -i\omega & \omega_0 \end{bmatrix}\begin{bmatrix} m_x \\ m_y \end{bmatrix}. \tag{3.60}$$

Inverting this equation yields the *Polder susceptibility tensor* :

$$\mathbf{m} = \overline{\overline{\chi}} \cdot \mathbf{h}, \tag{3.61}$$

where the *Polder susceptibility tensor* is given by

$$\overline{\overline{\chi}} = \begin{bmatrix} \chi & -i\kappa \\ i\kappa & \chi \end{bmatrix}, \tag{3.62}$$

and

$$\chi = \frac{\omega_0 \omega_M}{\omega_0^2 - \omega^2}, \tag{3.63a}$$

$$\kappa = \frac{\omega \omega_M}{\omega_0^2 - \omega^2}. \tag{3.63b}$$

The pole at ω_0 corresponds to the ferromagnetic *resonance frequency* in an infinite medium. The singularity can be removed by introducing energy dissipation, or damping.

To explore how the introduction of dissipation can be introduced, let us return to the original Landau–Lifshitz equation. Since dissipation will return the magnetization to its static equilibrium direction given sufficient time, we need a component of $d\mathbf{M}/dt$ in the $-\theta$ direction; i.e., in a direction that reduces the cone angle and drives the magnetization toward the equilibrium direction with $\mathbf{M}$ parallel to $\mathbf{H}$. Landau and Lifshitz [5] introduced a phenomenological term to do this proportional to $-(\mathbf{M} \times \mathbf{M} \times \mathbf{H}_{\text{eff}})$:

$$\frac{d\mathbf{M}}{dt} = \gamma\mu_0(\mathbf{M} \times \mathbf{H}_{\text{eff}}) + \frac{\lambda\gamma\mu_0}{M_S}\mathbf{M} \times (\mathbf{M} \times \mathbf{H}_{\text{eff}}), \tag{3.64}$$

while Gilbert suggested [6]

$$\frac{d\mathbf{M}}{dt} = \gamma\mu_0(\mathbf{M} \times \mathbf{H}_{\text{eff}}) + \frac{\alpha}{M_S}\left(\mathbf{M} \times \frac{d\mathbf{M}}{dt}\right). \tag{3.65}$$

The two forms are equivalent when λ and α are small (see Problem 3.8).

Let us consider the Gilbert form in more detail. Following the linearization procedure introduced earlier and again neglecting exchange and anisotropy, we obtain

$$i\omega\mathbf{m} = \hat{\mathbf{z}} \times [\omega_M\mathbf{h} - (\omega_0 - i\alpha\omega)\mathbf{m}]. \tag{3.66}$$

Comparing this with the lossless equation (3.57), we see that the effect of dissipation is simply to add a small imaginary part to the frequency ω_0.

Consequently damping can be added to the Polder tensor (3.62) by making the substitution $\omega_0 \to (\omega_0 - i\omega\alpha)$. For small enough α, $\mathbf{M}$ responds to an impulse by tilting away from $\hat{\mathbf{z}}$ at $t = 0$, and then slowly spiraling back to equilibrium, as described by the small-signal response:

$$m_x(t) \approx \omega_M e^{-\alpha\omega_0 t} \sin(\omega_0 t) u(t), \tag{3.67a}$$

$$m_y(t) \approx -\omega_M e^{-\alpha\omega_0 t} \cos(\omega_0 t) u(t). \tag{3.67b}$$

The eigenfunctions of (3.62) are also instructive in understanding the effect of introducing loss. Let us choose an excitation $\mathbf{h}_\pm = h_0 \mathbf{C}_\pm$, such that the magnetization is

$$
\begin{aligned}
\mathbf{m}_\pm &= \overline{\overline{\chi}} \cdot \mathbf{h}_\pm \\
&= h_0 \, \overline{\overline{\chi}} \cdot \mathbf{C}_\pm \\
&= h_0 \, \Lambda_\pm \mathbf{C}_\pm \\
\therefore \ \mathbf{m}_\pm &= \Lambda_\pm \mathbf{h}_\pm \equiv \chi_\pm \mathbf{h}_\pm ,
\end{aligned}
\tag{3.68}
$$

where we have made use of the eigenvalues

$$
\begin{aligned}
\chi_\pm &= \chi \pm \kappa \\
&= \frac{Z}{Z^2 - \Omega^2} \pm \frac{\Omega}{Z^2 - \Omega^2} \\
&= \frac{1}{Z \mp \Omega} .
\end{aligned}
\tag{3.69}
$$

The normalized eigenvectors corresponding to these eigenvalues are

$$
\mathbf{C}_\pm = \frac{1}{\sqrt{2}} \begin{bmatrix} 1 \\ \pm i \end{bmatrix} .
\tag{3.70}
$$

Here we have introduced the normalized quantities $Z = H/M_S$ and $\Omega = \omega/\omega_M$ to simplify the notation. We can now consider the behavior of the resonant susceptibility with loss:

$$
\begin{aligned}
\chi_+ &= \frac{1}{Z - i\Omega\alpha - \Omega} \\
&= \chi_+' + i\chi_+'' ,
\end{aligned}
\tag{3.71}
$$

with

$$
\chi_+' = \mathrm{Re}\{\chi_+\} = \frac{Z - \Omega}{(Z - \Omega)^2 + \Omega^2\alpha^2} ,
\tag{3.72}
$$

$$
\chi_+'' = \mathrm{Im}\{\chi_+\} = \frac{\Omega\alpha}{(Z - \Omega)^2 + \Omega^2\alpha^2} .
\tag{3.73}
$$

These functions are plotted in Fig. 3.3. The rate at which the magnetization returns to equilibrium is determined by the imaginary part of χ_+, and has the maximum value of $1/\Omega\alpha$ as the field Z is varied. The maximum occurs when $Z = \Omega$. The full width at half-maximum ΔZ is obtained as follows:

$$
\frac{\Omega\alpha}{(\Delta Z/2)^2 + \alpha^2\Omega^2} = \frac{1}{2\Omega\alpha} .
\tag{3.74}
$$

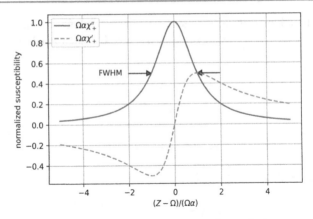

Fig. 3.3 Real and imaginary parts of the resonant susceptibility χ_+ in the presence of damping. (Adapted from [2] with permission)

Solving for ΔZ gives

$$\Delta Z = 2\Omega\alpha, \tag{3.75a}$$

$$\Delta B = -\frac{2\omega\alpha}{\gamma}, \tag{3.75b}$$

or

$$\Delta H = -\frac{2\omega\alpha}{\gamma\mu_0}, \tag{3.75c}$$

where ΔZ, ΔB, and ΔH correspond to the full resonance line width at half-maximum.

Recalling (3.67), the amplitude of the small-signal magnetization is of the form

$$|\mathbf{m}| = m_0 e^{-\alpha\omega_0 t}. \tag{3.76}$$

Using the expression for the full width at half-maximum (3.75b) and recognizing that $\omega = \omega_0$ at resonance gives

$$|\mathbf{m}| = m_0 e^{\frac{\gamma\Delta B t}{2}}. \tag{3.77}$$

The normalized magnitude expressed in dB is

$$\frac{|\mathbf{m}|}{m_0}(\text{dB}) = 20\log e^{\frac{\gamma\Delta B t}{2}}$$

or,

$$\frac{|\mathbf{m}|}{m_0}(\text{dB}) = -76.4 \times 10^{10}\Delta B t, \tag{3.78}$$

where ΔB is in Teslas and t is in seconds. We define the loss parameter L as the attenuation in dB/ns[3]:

$$L = 764\Delta B \text{ (dB/ns)}, \tag{3.79}$$

where ΔB is again given in Teslas.

Problem 3.8 Find γ' and the relationship between λ and α that allow the Landau–Lifshitz equation to be cast into the same form as the Landau–Lifshitz–Gilbert equation even if α and λ are not small:

$$LL: \quad \frac{d\mathbf{M}}{dt} = \gamma'\mu_0(\mathbf{M} \times \mathbf{H}_{\text{eff}}) + \frac{\lambda\gamma'\mu_0}{M_S}\mathbf{M} \times (\mathbf{M} \times \mathbf{H}_{\text{eff}}), \tag{3.80}$$

$$LLG: \quad \frac{d\mathbf{M}}{dt} = \gamma\mu_0(\mathbf{M} \times \mathbf{H}_{\text{eff}}) + \frac{\alpha}{M_S}\left(\mathbf{M} \times \frac{d\mathbf{M}}{dt}\right). \tag{3.81}$$

Although the equations can be cast into the same form, they describe different physics since the dependence on the loss parameter differs between the two cases (e.g., consider what happens when loss becomes large in both cases [7]).

Solution 3.8

We need to find γ' and the relationship between λ and α such that the Landau–Lifshitz equation can be cast into the same form as the Landau–Lifshitz–Gilbert equation even if α and λ are not small:

$$LL: \quad \frac{d\overline{M}}{dt} = \gamma'\mu_0\overline{M} \times \overline{H} + \frac{\lambda\gamma'\mu_0}{M}\overline{M} \times (\overline{M} \times \overline{H}) \tag{3.82}$$

$$LLG: \quad \frac{d\overline{M}}{dt} = \gamma\mu_0\overline{M} \times \overline{H} + \frac{\alpha}{M}\overline{M} \times \frac{d\overline{M}}{dt} \tag{3.83}$$

Taking the cross product $\overline{M} \times$ (3.82),

$$\overline{M} \times \frac{d\overline{M}}{dt} = \gamma'\mu_0\overline{M} \times \left[\overline{M} \times \overline{H} + \frac{\lambda}{M}\left((\overline{M} \cdot \overline{H})\overline{M} - M^2\overline{H}\right)\right]$$

$$= \gamma'\mu_0\overline{M} \times (\overline{M} \times \overline{H}) - \gamma'\mu_0 M\lambda \overline{M} \times \overline{H}$$

$$= \left(\frac{M}{\lambda}\right)\gamma'\frac{\mu_0\lambda}{M}\overline{M} \times (\overline{M} \times \overline{H}) + \left(\frac{M}{\lambda}\right)\gamma'\mu_0\overline{M} \times \overline{H}$$

$$- \left(\frac{M}{\lambda}\right)\gamma'\mu_0\overline{M} \times \overline{H} - \gamma'\mu_0 M\lambda \overline{M} \times \overline{H}$$

$$= \frac{M}{\lambda}\frac{d\overline{M}}{dt} - \gamma'\mu_0 M\left(\frac{1}{\lambda} + \lambda\right)\overline{M} \times \overline{H}$$

---

[3]The analogous cgs expression is $L = 76.4 \, \Delta H$ (dB/μs), with ΔH given in Oe.

Rearranging the terms on boths sides of the equation, we get the LLG form

$$\frac{d\overline{M}}{dt} = \gamma' \mu_0 \left(1 + \lambda^2\right) \overline{M} \times \overline{H} + \frac{\lambda}{M} \overline{M} \times \frac{d\overline{M}}{dt}$$

A comparison with (3.83) suggests

$$\boxed{\gamma' = \frac{\gamma}{\left(1 + \lambda^2\right)}, \ \text{and} \ \alpha = \lambda.}$$

Although the equations can be cast into the same form, they describe different physics since γ does not depend on λ whereas γ' does. However, for $\alpha = \lambda \ll 1$, the forms are equivalent to a good approximation.

Problem 3.9 A low-loss ferrite is found to have a ferromagnetic resonance (FMR) line width of $\Delta B = 50\,\mu T$ at $9.3\,GHz$. If the external field driving the resonance is removed at $t=0$, how long will it take for the small-signal magnetization to decay by $10\,dB$ relative to its magnitude at $t = 0$?

Solution 3.9

$$\Delta B = 50\,\mu T \ \text{at} \ 9.3\,GHz$$
$$L = 764\,\Delta B \ dB/ns$$
$$= 764(50)10^{-6}$$
$$= 0.0382\,dB/ns$$

Hence
$$\boxed{\frac{10\,dB}{0.0382\,dB/ns} = 262\,ns.}$$

Problem 3.10 Most ferromagnetic resonance spectrometers measure the derivative of the absorption with respect to the bias field H_0 rather than the absorption curve itself. The peak-to-peak resonance line width ΔH_{pp} is obtained from this curve. Find the separation between the peaks of the derivative of the imaginary part of the susceptibility

$$\chi''_+ = Im\{\chi_+\} = \frac{\Omega \alpha}{(Z - \Omega)^2 + \Omega^2 \alpha^2}, \tag{3.84}$$

and show that $\Delta H_{FWHM} = \sqrt{3}\,\Delta H_{pp}$, where ΔH_{FWHM} is the full width at half-maximum of the absorption curve and is given by

$$\Delta H = -\frac{2\omega \alpha}{\gamma \mu_0}. \tag{3.85}$$

Solution 3.10

From (3.73), we have

$$\chi''_+ = \frac{\Omega\alpha}{(Z-\Omega)^2 - \Omega^2\alpha^2}$$

with the derivatives

$$\frac{\partial\chi''_+}{\partial Z} = \frac{-2\Omega\alpha(Z-\Omega)}{[(Z-\Omega)^2 + \Omega^2\alpha^2]^2},$$

$$\frac{\partial^2\chi''_+}{\partial Z^2} = \frac{-2\Omega\alpha}{[(Z-\Omega)^2 + \Omega^2\alpha^2]^2} + \frac{8\Omega\alpha(Z-\Omega)^2}{[(Z-\Omega)^2 + \Omega^2\alpha^2]^3}.$$

The peaks in the first derivative occur when the second derivative vanishes.

$$\frac{\partial^2\chi''_+}{\partial Z^2} = 0 \Rightarrow \frac{4(Z-\Omega)^2}{(Z-\Omega)^2 + \Omega^2\alpha^2} = 1.$$

$$\therefore (Z-\Omega)^2 + \Omega^2\alpha^2 = 4(Z-\Omega)^2$$
$$\Omega^2\alpha^2 = 3(Z-\Omega)^2$$
$$\pm\Omega\alpha = \sqrt{3}(Z-\Omega)$$
$$\Rightarrow \quad Z = \Omega \mp \frac{\Omega\alpha}{\sqrt{3}}.$$

The distance between the peaks is

$$\Delta Z_{pp} = \frac{2\Omega\alpha}{\sqrt{3}} = \frac{\Delta Z_{FWHM}}{\sqrt{3}}.$$

Thus, $\dfrac{\Delta H_{pp}}{M_S} = \dfrac{\Delta H_{FWHM}}{M_S\sqrt{3}}$ and

$$\boxed{\Delta H_{FWHM} = \sqrt{3}\,\Delta H_{pp}.}$$

3.7 Stoner–Wohlfarth Model

A popular quasistatic model for the switching of magnetic particles was introduced by Stoner and Wohlfarth [8]. In this model we assume a uniformly magnetized particle in the shape of a prolate ellipsoid, as shown in Fig. 3.4a. The orientation of the magnetization **M** is determined by minimizing the total energy of the particle,

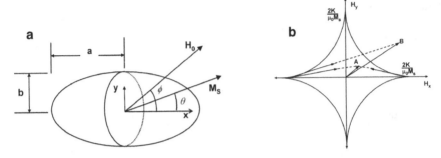

Fig. 3.4 a Single domain prolate ellipsoid with a field **H** applied at an angle ϕ with respect to the major axis. (Reproduced from [2] with permission.) **b** The Stoner–Wohlfarth asteroid. Tangents to the asteroid (in the upper half-plane) are used to determine the stability of the resulting magnetic orientation. (Reproduced from [2] with permission)

and the stability of the solution is determined using the asteroid curve shown in Fig. 3.4b.

Assume that the particle has a saturation magnetization M_S, and its major axis (its "easy axis") is along $\hat{x}$. If an external field H_0 is applied at an angle ϕ relative to $\hat{x}$, the magnetization of the particle will equilibrate at an angle θ relative to $\hat{x}$. The total energy is the sum of Zeeman and anisotropy contributions:

$$W = -\mu_0 M_S H_x \cos\theta - \mu_0 M_S H_y \sin\theta + K \sin^2\theta, \tag{3.86}$$

where $H_x = H_0 \cos\phi$, $H_y = H_0 \sin\phi$, and we assume an anisotropy constant K (c.f. (3.33)). Defining the anisotropy field $H_k = \frac{2K}{\mu_0 M_S}$, we find the condition for the particle to orient itself at an energy minimum:

$$\frac{\partial W}{\partial \theta} = 0 \quad \Rightarrow \quad H_x \sin\theta - H_y \cos\theta + H_k \sin\theta \cos\theta = 0. \tag{3.87}$$

However, stability is ensured only when the second derivative, evaluated at angles θ_0 that satisfy (3.87), also satisfies $\frac{\partial^2 W}{\partial \theta^2}\Big|_{\theta_0} > 0$. The boundary of stability is easily determined as

$$\frac{\partial^2 W}{\partial \theta^2}\Big|_{\theta_0} = 0 \quad \Rightarrow \quad H_x \cos\theta + H_y \sin\theta + H_k \cos 2\theta = 0. \tag{3.88}$$

Combining (3.87) and (3.88) we get $H_x = -H_k \cos^3\theta$ and $H_y = H_k \sin^3\theta$, that together form the Stoner–Wohlfarth *asteroid*, shown in Fig. 3.4b,

$$H_x^{2/3} + H_y^{2/3} = H_k^{2/3}. \tag{3.89}$$

While the Stoner–Wohlfarth model allows us to determine the final orientation of M_S it does not help us determine the trajectory of the magnetization as the particle reaches its minimum energy. The magnetization dynamics continue to be governed by the Landau–Lifshitz–Gilbert equation.

Problem 3.11 A field $\mathbf{H} = H_a\,\hat{\mathbf{x}} + H_b\,\hat{\mathbf{y}}$ is applied to a Stoner–Wohlfarth particle. Referring to Fig. 3.4b, we observe that the magnetization of the particle is stable when it makes an angle θ with respect to the x-axis. Show that a line with slope $\tan\theta$, tangential to the asteroid at a point (H_1, H_2) and passing through the point (H_a, H_b), satisfies the equation:

$$H_a \sin\theta - H_b \cos\theta + H_k \sin\theta \cos\theta = 0. \tag{3.90}$$

Solution 3.11

The slope of a line going through (H_1, H_2) and (H_a, H_b) is

$$\tan\theta = (H_b - H_2)/(H_a - H_1). \tag{3.91}$$

Since (H_1, H_2) lies on the asteroid, it must satisfy

$$H_1 = -H_k \sin^3\theta, \tag{3.92}$$
$$H_2 = H_k \cos^3\theta. \tag{3.93}$$

Substituting H_1 and H_2 into (3.91), and using the identity

$$\sin^2\theta + \cos^2\theta = 1,$$

we obtain

$$\boxed{H_a \sin\theta - H_b \cos\theta + H_k \sin\theta \cos\theta = 0.} \tag{3.94}$$

References

1. R. Kubo, T. Nagamiya, *Solid State Physics* (McGraw-Hill, New York, 1969)
2. D.D. Stancil, A. Prabhakar, *Spin Waves: Theory and Applications* (Springer, New York, 2009)
3. C. Kittel, *Introduction to Solid State Physics*, 8th edn. (Wiley, New York, 2005)
4. C. Vittoria, G.C. Bailey, R.C. Barker, A. Yelon, Ferromagnetic resonance field and linewidth in an anisotropic magnetic metallic medium. Phys. Rev. B **7**, 2112 (1973)
5. L. Landau, L. Lifshitz, On the theory of the dispersion of magnetic permeability in ferromagnetic bodies. Phys. Zeit. Sowjetunion **8**, 153 (1935)
6. T.A. Gilbert, Equation of motion of magnetization, Armor Research Foundation, Chicago, IL. Technical Report (1995).
7. J.C. Mallinson, On damped gyromagnetic precession. IEEE Trans. Mag. **MAG-23**(4), 2003 (1987)
8. E.C. Stoner, E.P. Wohlfarth, A mechanism of magnetic hysteresis in heterogeneous alloys. *Philos. Trans. R. Soc. B, Mathematical and Physical Sciences*, **240**, 599–642 (1948)

Electromagnetic Waves in Anisotropic Dispersive Media

4

Electromagnetic waves traveling in media often elicit a response from the medium that depends both on the direction of propagation of the wave and the direction of the electromagnetic field. This behavior is captured by the permeability and permittivity tensors. In Problems 4.1 and 4.2, we introduce the concepts of energy flux, energy densities, the Poynting vector, and phase and group velocities.

The two wave solutions to Maxwell's equations in anisotropic media are called the *ordinary* wave and the *extraordinary* wave. We describe wave propagation in a uniaxial crystal in Problem 4.3 and calculate and plot the dispersion relations for circularly polarized modes propagating in a ferrite in Problem 4.4. Problem 4.5 introduces the concept of *Faraday rotation* and describes how wave propagation in a ferrite medium is non-reciprocal, and depends on the direction of propagation relative to an applied field. Propagation in a ferrite is further explored in Problem 4.6 where we consider plane wave propagation perpendicular to the applied field.

4.1 Maxwell's Equations

To understand electromagnetic waves, we begin with *Maxwell's Equations*[1]:

$$\nabla \times \mathbf{E} = -\frac{\partial \mathbf{B}}{\partial t}, \tag{4.1a}$$

$$\nabla \times \mathbf{H} = \frac{\partial \mathbf{D}}{\partial t} + \mathbf{J}, \tag{4.1b}$$

$$\nabla \cdot \mathbf{B} = 0, \tag{4.1c}$$

$$\nabla \cdot \mathbf{D} = \rho, \tag{4.1d}$$

---

[1] A standard textbook such as Ulaby [1] is a good source of introductory material on Maxwell's equations.

where $\mathbf{E}$ is the electric field intensity (V/m), $\mathbf{B}$ is the magnetic flux density (T), $\mathbf{H}$ is the magnetic field intensity (A/m), $\mathbf{D}$ is the electric flux density (C/m^2), $\mathbf{J}$ is the electric current density (A/m^2), and ρ is the electric charge density (C/m^3). In a classical physics context (i.e., without quantization of the electromagnetic field), the equations in the form given are exact and contain no approximations.

When studying electromagnetic waves it is often more convenient to use *phasor* notation to represent the fields. For example, a sinusoidally-varying field at frequency ω would be written $\mathbf{E}(\omega) \exp(-i\omega t)$, where $\mathbf{E}(\omega)$ is a complex amplitude that in general is frequency dependent. We will use the same notation for the real time domain amplitude and the complex phasor amplitude. When the context does not make it clear, we will explicitly show the dependence on either ω or t. The time domain expression is obtained by taking the real part of the phasor:

$$\mathbf{E}(t) = \mathrm{Re}\left(\mathbf{E}(\omega)e^{-i\omega t}\right). \tag{4.2}$$

With this form, the time derivatives can simply be replaced with $-i\omega$. Maxwell's Equations then take the form

$$\nabla \times \mathbf{E} = i\omega \mathbf{B}, \tag{4.3a}$$

$$\nabla \times \mathbf{H} = -i\omega \mathbf{D} + \mathbf{J}, \tag{4.3b}$$

$$\nabla \cdot \mathbf{B} = 0, \tag{4.3c}$$

$$\nabla \cdot \mathbf{D} = \rho. \tag{4.3d}$$

In a similar way, phasor notation can be used to represent an electromagnetic plane wave as follows:

$$\mathbf{E}(\mathbf{r}, t) = \mathrm{Re}\left\{\mathbf{E}(\omega, \mathbf{k})e^{-i\omega t + i\mathbf{k} \cdot \mathbf{r}}\right\}, \tag{4.4}$$

where $\mathbf{r} = (x, y, z)$ and $\mathbf{k} = (k_x, k_y, k_z)$. When the fields are of this form, the del operator can be replaced by the factor $i\mathbf{k}$. Maxwell's equations for plane waves then take the form:

$$\mathbf{k} \times \mathbf{E} = \omega \mathbf{B}, \tag{4.5a}$$

$$\mathbf{k} \times \mathbf{H} = -\omega \mathbf{D} - i\mathbf{J}, \tag{4.5b}$$

$$\mathbf{k} \cdot \mathbf{B} = 0, \tag{4.5c}$$

$$i\mathbf{k} \cdot \mathbf{D} = \rho. \tag{4.5d}$$

4.2 Constitutive Relations[2]

However, Maxwell's equations by themselves do not provide enough information to solve for the field quantities. What is needed is information about the medium in which the fields exist. This information is provided by *constitutive relations*. For linear media, the constitutive relations can be written

$$\mathbf{D} = \bar{\varepsilon} \cdot \mathbf{E}, \tag{4.6}$$

$$\mathbf{B} = \bar{\mu} \cdot \mathbf{H}, \tag{4.7}$$

$$\mathbf{J} = \bar{\sigma} \cdot \mathbf{E}. \tag{4.8}$$

The quantities $\bar{\varepsilon}$, $\bar{\mu}$, and $\bar{\sigma}$ are the *permittivity tensor, permeability tensor* and *conductivity tensor*, respectively.[3] We will focus our attention on insulators ($\bar{\sigma} = \bar{0}$) and so will not consider the conductivity tensor, current density $\mathbf{J}$, or charge density ρ further.

Generally speaking, the electromagnetic properties of a material must be represented by tensors if the properties of the material vary with the direction of the applied fields, or if application of a field along one direction causes a material response in a different direction. Once a coordinate system is selected, these tensors can be represented by 3×3 matrices. If a coordinate system can be found in which the matrix representing the tensor is diagonal, the material is called *biaxial* if all three diagonal elements are different:

$$\begin{bmatrix} \varepsilon_x & 0 & 0 \\ 0 & \varepsilon_y & 0 \\ 0 & 0 & \varepsilon_z \end{bmatrix}, \tag{4.9}$$

uniaxial if two diagonal elements are equal and different from the third, e.g.:

$$\begin{bmatrix} \varepsilon & 0 & 0 \\ 0 & \varepsilon & 0 \\ 0 & 0 & \varepsilon_z \end{bmatrix}, \tag{4.10}$$

and *isotropic* if all three diagonal elements are equal:

$$\begin{bmatrix} \varepsilon & 0 & 0 \\ 0 & \varepsilon & 0 \\ 0 & 0 & \varepsilon \end{bmatrix} \equiv \varepsilon \bar{\mathbf{I}}. \tag{4.11}$$

---

[2]Discussions of constitutive relations for anisotropic and bianisotropic media can be found in Chen [2], Haus [3], and Kong [4].

[3]In general, $\mathbf{D}$ and $\mathbf{B}$ can each depend on both $\mathbf{E}$ and $\mathbf{H}$. In such materials, application of a magnetic field will induce an electric polarization, and application of an electric field will induce a magnetic moment. Materials with this property are referred to as *magnetoelectric*. If the coefficients are tensors, magnetoelectric media are also referred to as *bianisotropic*.

A tensor with complex off-diagonal elements is *Hermitian* if the transpose complex conjugate is the same as the original tensor. Such tensors with imaginary off-diagonal elements are referred to as *gyrotropic*, e.g.,

$$\begin{bmatrix} \mu_x & -i\mu_g & 0 \\ i\mu_g & \mu_y & 0 \\ 0 & 0 & \mu_z \end{bmatrix}. \tag{4.12}$$

Ferro- or ferri-magnetic materials magnetized along the z axis have permeabilities of this form. Specifically,

$$\overline{\mu} = \mu_0 \begin{bmatrix} 1+\chi & -i\kappa & 0 \\ i\kappa & 1+\chi & 0 \\ 0 & 0 & 1 \end{bmatrix}, \tag{4.13}$$

and χ and κ are given by (3.63a) and (3.63b), respectively.

4.3   Time Averages and Energy Conservation

The time average value of the product of two quantities expressed in the time domain is given by

$$\langle a(t)b(t) \rangle = \frac{1}{T} \int_T a(t)b(t)\mathrm{d}t. \tag{4.14}$$

For sinusoidally-varying quantities, the time average is obtained conveniently from the phasor or frequency-domain expressions by taking one half the real part of the product of one times the complex conjugate of the other:

$$\langle a(t)b(t) \rangle = \frac{1}{2}\mathrm{Re}\{a^*b\}. \tag{4.15}$$

We can use this property to calculate useful results concerning energy conservation from Maxwell's equations. Specifically, let us dot multiply the complex conjugate of (4.3b) by $\mathbf{E}$, dot multiply (4.3a) by $\mathbf{H}^*$, and take the difference to obtain

$$\mathbf{H}^* \cdot (\nabla \times \mathbf{E}) - \mathbf{E} \cdot (\nabla \times \mathbf{H}^*) + i\omega\mathbf{E} \cdot \mathbf{D}^* - i\omega\mathbf{H}^* \cdot \mathbf{B} = 0 \tag{4.16}$$

where we have taken $\mathbf{J} = \mathbf{0}$ for an insulator. This expression can be simplified using the vector identity

$$\nabla \cdot (\mathbf{A} \times \mathbf{B}) = \mathbf{B} \cdot (\nabla \times \mathbf{A}) - \mathbf{A} \cdot (\nabla \times \mathbf{B}). \tag{4.17}$$

The result is

$$\nabla \cdot (\mathbf{E} \times \mathbf{H}^*) + i\omega(\mathbf{E} \cdot \mathbf{D}^* - \mathbf{H}^* \cdot \mathbf{B}) = 0. \tag{4.18}$$

Recognizing that taking the real part of each term can be related to the time average from (4.15), we identify the following quantities:

$$\langle \mathbf{P}(t) \rangle = \frac{1}{2} \mathrm{Re} \left\{ \mathbf{E} \times \mathbf{H}^* \right\}, \tag{4.19}$$

$$\langle w_e(t) \rangle = \frac{1}{2} \mathrm{Re} \left\{ \mathbf{E} \cdot \mathbf{D}^* \right\}, \tag{4.20}$$

$$\langle w_m(t) \rangle = \frac{1}{2} \mathrm{Re} \left\{ \mathbf{H}^* \cdot \mathbf{B} \right\}, \tag{4.21}$$

where $\langle \mathbf{P}(t) \rangle$ is the time-averaged *Poynting vector*, and represents the energy per unit time crossing a unit area of a surface perpendicular to $\mathbf{P}$, $\langle w_e(t) \rangle$ is the time-averaged energy stored in the electric field, and $\langle w_m(t) \rangle$ is the time-averaged energy stored in the magnetic field.

When we integrate (4.18) over a sample volume V, then the first term will yield the net power leaving the volume during one period of the oscillation. In the case of a lossless material, all the power that enters (on average) must also leave the volume. Consequently, we will have

$$\mathrm{Re}\{i\omega \left(\mathbf{E} \cdot \mathbf{D}^* - \mathbf{H}^* \cdot \mathbf{B} \right)\} = 0. \tag{4.22}$$

Problem 4.1 Show that

$$\mathrm{Re}\{i\omega \mathbf{E} \cdot \mathbf{D}^*\} = \frac{i\omega}{2} \mathbf{E}^* \cdot (\overline{\overline{\varepsilon}}^\dagger - \overline{\overline{\varepsilon}}) \cdot \mathbf{E}, \tag{4.23}$$

and

$$\mathrm{Re}\{-i\omega \mathbf{H}^* \cdot \mathbf{B}\} = \frac{i\omega}{2} \mathbf{H}^* \cdot (\overline{\overline{\mu}}^\dagger - \overline{\overline{\mu}}) \cdot \mathbf{H}, \tag{4.24}$$

and the lossless condition (4.22) will be satisfied when

$$\overline{\overline{\varepsilon}}^\dagger = \overline{\overline{\varepsilon}} \tag{4.25}$$

$$\overline{\overline{\mu}}^\dagger = \overline{\overline{\mu}}. \tag{4.26}$$

Solution 4.1

$$\mathrm{Re}\left\{ i\omega \overline{E} \cdot \overline{D}^* \right\} = \frac{i\omega}{2} \left[\overline{E} \cdot \overline{D}^* - E^* \cdot D \right]$$

$$= \frac{i\omega}{2} \left[\overline{E} \cdot \left(\overline{\overline{\varepsilon}}^* \cdot \overline{E}^* \right) - \overline{E}^* \cdot \overline{\overline{\varepsilon}} \cdot \overline{E} \right]$$

$$= \frac{i\omega}{2} \left[\overline{E}^* \cdot \overline{\overline{\varepsilon}}^\dagger \cdot \overline{E} - \overline{E}^* \cdot \overline{\overline{\varepsilon}} \cdot \overline{E} \right]$$

(Recall that $\left(\overline{\overline{A}} \cdot \overline{\overline{B}} \cdot \overline{\overline{C}} \right)^{\mathrm{T}} = \overline{\overline{C}}^{\mathrm{T}} \cdot \overline{\overline{B}}^{\mathrm{T}} \cdot \overline{\overline{A}}^{\mathrm{T}}$, and $a^{\mathrm{T}} = a$, if a is a scalar.)

$$= \frac{i\omega}{2} \quad \overline{E}^* \cdot \left(\overline{\overline{\varepsilon}}^\dagger - \overline{\overline{\varepsilon}} \right) \cdot \overline{E}$$

$$\therefore \; \mathrm{Re} \left\{ i\omega \overline{E} \cdot \overline{D}^* \right\} = \frac{i\omega}{2} \; \overline{E}^* \cdot \left(\overline{\overline{\varepsilon}}^\dagger - \overline{\overline{\varepsilon}} \right) \cdot \overline{E}.$$

Similarly

$$\mathrm{Re} \left\{ -i\omega \overline{H}^* \cdot \overline{B} \right\} = -\frac{i\omega}{2} \left[\overline{H}^* \cdot \overline{B} - \overline{H} \cdot \overline{B}^* \right]$$

$$= \frac{-i\omega}{2} \left[\overline{H}^* \cdot \overline{\overline{\mu}} \cdot \overline{H} - \overline{H} \cdot \overline{\overline{\mu}}^* \cdot \overline{H}^* \right]$$

$$= \frac{-i\omega}{2} \left[\overline{H}^* \cdot \overline{\overline{\mu}} \cdot \overline{H} - \overline{H}^* \cdot \overline{\overline{\mu}}^\dagger \cdot \overline{H} \right]$$

$$= \frac{i\omega}{2} \overline{H}^* \cdot \left(\overline{\overline{\mu}}^\dagger - \overline{\overline{\mu}} \right) \cdot \overline{H}$$

$$\therefore \; \mathrm{Re} \left\{ -i\omega \overline{H}^* \cdot \overline{B} \right\} = \frac{i\omega}{2} \overline{H}^* \cdot \left(\overline{\overline{\mu}}^\dagger - \overline{\overline{\mu}} \right) \cdot \overline{H}.$$

When the permittivity and permeability matrices are Hermitian, i.e., $\overline{\overline{\varepsilon}}^\dagger = \overline{\overline{\varepsilon}}$ and $\overline{\overline{\mu}}^\dagger = \overline{\overline{\mu}}$,

$$\mathrm{Re}\{i\omega \, (\overline{E} \cdot \overline{D}^* - \overline{H}^* \cdot \overline{B})\} = \frac{i\omega}{2} \left[\overline{E}^* \cdot (\overline{\overline{\varepsilon}}^\dagger - \overline{\overline{\varepsilon}}) \cdot \overline{E} + \overline{H}^* \cdot (\overline{\overline{\mu}}^\dagger - \overline{\overline{\mu}}) \cdot \overline{H} \right]$$

$$= 0,$$

and the medium is lossless.

4.4   Plane Waves

Consider a solution to Maxwell's Equations of the form (4.4). The phase shift with position is given by the second term in the exponential: $\phi = \mathbf{k} \cdot \mathbf{r}$. If we set $\phi = $ constant, this is the equation for a plane surface perpendicular to $\mathbf{k}$. Since the constant-phase surface is a plane and the amplitude is uniform along the constant-phase surface, a solution of this form is called a *uniform plane wave*. As time increases, this plane moves in the direction of $\mathbf{k}$ with speed ω/k. We therefore define the *phase velocity* as

$$\mathbf{v}_p = \frac{\omega}{k} \hat{\mathbf{k}}. \tag{4.27}$$

To aid in the manipulation of Maxwell's Equations (4.5) it is convenient to define the matrix $\overline{\mathbf{k}}$ as

$$\overline{\mathbf{k}} \equiv \mathbf{k} \times \overline{\mathbf{I}} = \begin{bmatrix} 0 & -k_z & k_y \\ k_z & 0 & -k_x \\ -k_y & k_x & 0 \end{bmatrix}. \tag{4.28}$$

It is easy to verify that this matrix has the following property:

$$\overline{\mathbf{k}} \cdot \mathbf{A} = \mathbf{k} \times \mathbf{A} \tag{4.29}$$

for an arbitrary vector $\mathbf{A}$. Using this notation along with the constitutive relations (4.6) and (4.7), Maxwell's Equations from Faraday's and Ampere's laws (4.5a) and (4.5b) in a source-free region become

$$\overline{\mathbf{k}} \cdot \mathbf{E} = \omega \overline{\mu} \cdot \mathbf{H}, \tag{4.30a}$$

$$\overline{\mathbf{k}} \cdot \mathbf{H} = -\omega \overline{\varepsilon} \cdot \mathbf{E}. \tag{4.30b}$$

Multiplying the first equation from the left by the inverse of the permeability tensor and dotting $\overline{\mathbf{k}}$ into both sides from the left gives

$$\overline{\mathbf{k}} \cdot \overline{\mu}^{-1} \cdot \overline{\mathbf{k}} \cdot \mathbf{E} = \omega \overline{\mathbf{k}} \cdot \mathbf{H}. \tag{4.31}$$

Finally, $\overline{\mathbf{k}} \cdot \mathbf{H}$ can be eliminated by substituting (4.30b) to obtain a single equation for $\mathbf{E}$:

$$\left[\overline{\mathbf{k}} \cdot \overline{\mu}^{-1} \cdot \overline{\mathbf{k}} + \omega^2 \overline{\varepsilon} \right] \cdot \mathbf{E} = 0. \tag{4.32}$$

Following a similar procedure to eliminate $\mathbf{E}$ gives

$$\left[\overline{\mathbf{k}} \cdot \overline{\varepsilon}^{-1} \cdot \overline{\mathbf{k}} + \omega^2 \overline{\mu} \right] \cdot \mathbf{H} = 0. \tag{4.33}$$

Equations (4.32) and (4.33) are *complex wave equations* for $\mathbf{E}$ and $\mathbf{H}$, respectively. Setting the determinants of the coefficient matrices to zero give the possible relations between ω and $\mathbf{k}$ that will permit a non-trivial solution for $\mathbf{E}$. Both equations are equivalent, but one may be mathematically more convenient than the other depending on how hard it is to invert the permittivity or the permeability tensor.

If either the permittivity or permeability is isotropic, then the quantity $\overline{\mathbf{k}} \cdot \overline{\mathbf{k}}$ appears in the dispersion equations (4.32) and (4.33). In such cases it is often possible to simplify the equation using the identity

$$\overline{\mathbf{k}} \cdot \overline{\mathbf{k}} = \mathbf{k}\mathbf{k} - k^2 \overline{\mathbf{I}}, \tag{4.34}$$

where the *dyadic product* of two vectors $\mathbf{a}$ and $\mathbf{b}$ is given by $[\mathbf{ab}]_{ij} = a_i b_j$.

Problem 4.2 The electric field of a plane wave propagating in free space is given by

$$\mathbf{E} = \hat{\mathbf{y}} E_0 e^{ikz - i\omega t}. \tag{4.35}$$

(a) Using Maxwell's equations, find $\mathbf{H}$.
(b) Calculate the time-averaged electric and magnetic energy densities and show that

$$\langle w_m(t) \rangle = \langle w_e(t) \rangle. \tag{4.36}$$

(c) Calculate the energy velocity defined by

$$\mathbf{v_e} \equiv \frac{\langle \mathbf{P}(t) \rangle}{\langle w(t) \rangle} \tag{4.37}$$

and show that

$$\mathbf{v_e} = \mathbf{v_p} = \frac{\omega}{k}\hat{\mathbf{z}}. \tag{4.38}$$

(d) Find an expression for the *intrinsic wave impedance* of the medium defined as

$$Z_0 \equiv \frac{|\mathbf{E}|}{|\mathbf{H}|}. \tag{4.39}$$

Use the dispersion relation to express this in a form independent of ω and k.

Solution 4.2

We are given the form $\overline{E} = \hat{y}E_0 e^{ikz-i\omega t}$

(a) Starting from Maxwell's equation:

$$\nabla \times \overline{E} = i\omega\mu_0\overline{H}$$
$$\overline{k} \times \overline{E} = \omega\mu_0\overline{H}$$

$$\boxed{\therefore \ \overline{H} = \frac{k}{\omega\mu_0}\hat{z} \times \overline{E} = \frac{-kE_0}{\omega\mu_0}\hat{x}e^{ikz-i\omega t}}$$

(b) The time-averaged energy terms are

$$\langle w_e(t) \rangle = \frac{1}{4}\varepsilon_0\overline{E}^* \cdot \overline{E} = \frac{\varepsilon_0}{4}E_0^2$$

$$\langle w_m(t) \rangle = \frac{\mu_0}{4}\overline{H}^* \cdot \overline{H} = \frac{\mu_0}{4}\left(\frac{kE_0}{\omega\mu_0}\right)^2.$$

But, we know that $k = \omega\sqrt{\mu_0\varepsilon_0}$. Therefore,

$$\langle w_m(t) \rangle = \frac{\mu_0}{4}\left(\sqrt{\frac{\varepsilon_0}{\mu_0}}E_0\right)^2 = \frac{\varepsilon_0}{4}E_0^2, \quad \text{and} \quad \boxed{\langle w_m(t) \rangle = \langle w_e(t) \rangle}.$$

(c) The energy velocity is defined as

$$v_e = \frac{\langle \overline{P}(t) \rangle}{\langle w(t) \rangle}.$$

Now, the time-averaged Poynting vector is

$$\langle \overline{P}(t) \rangle = \frac{1}{2} \overline{E} \times \overline{H}^* = \frac{1}{2} \frac{k E_0^2}{\omega \mu_0} \left(\hat{y} \times (-\hat{x}) \right) = \frac{1}{2} \frac{k E_0^2}{\omega \mu_0} \hat{z},$$

while from Part (b) we have

$$\langle w(t) \rangle = \langle w_e(t) \rangle + \langle w_m(t) \rangle = \frac{\varepsilon_0}{2} E_0^2.$$

Hence,

$$\overline{v}_e = \frac{\frac{1}{2} \frac{k E_0^2}{\omega \mu_0}}{\frac{\varepsilon_0}{2} E_0^2} \hat{z} = \frac{k}{\omega \mu_0 \varepsilon_0} \hat{z} = \frac{\omega \sqrt{\mu_0 \varepsilon_0}}{(\omega \sqrt{\mu_0 \varepsilon_0}) \sqrt{\mu_0 \varepsilon_0}} \hat{z} = \frac{\omega}{k} \hat{z}$$

Therefore,

$$\boxed{\overline{v}_e = \frac{\omega}{k} \hat{z} \equiv \overline{v}_p.}$$

(d) The intrinsic wave impedance is

$$Z_0 = \frac{|\overline{E}|}{|\overline{H}|} = \frac{E_0}{\frac{k E_0}{\omega \mu_0}} = \frac{\omega \mu_0}{k}.$$

Substituting the dispersion relation $\omega = ck$ in the denominator, we recover the characteristic impedance of free space:

$$\boxed{Z_0 = \frac{k \mu_0}{k \sqrt{\mu_0 \varepsilon_0}} = \sqrt{\frac{\mu_0}{\varepsilon_0}} \approx 377 \, \Omega.}$$

Problem 4.3 The permittivity tensor of a uniaxial dielectric is of the form

$$\overline{\varepsilon} = \begin{bmatrix} \varepsilon & 0 & 0 \\ 0 & \varepsilon & 0 \\ 0 & 0 & \varepsilon_z \end{bmatrix}. \tag{4.40}$$

The $\hat{z}$ direction is called the *optic axis*.

(a) Assuming uniform plane wave propagation and a nonmagnetic medium, show that the electric field must satisfy the equation

$$\overline{W}(\omega, \mathbf{k}) \cdot \mathbf{E} = 0, \tag{4.41}$$

where $\overline{W}(\omega, \mathbf{k})$ is a 3×3 matrix whose elements are functions of $(\omega, \mathbf{k})$. $\overline{W}(\omega, \mathbf{k})$ is sometimes called the *wave matrix* or *dispersion matrix*.

(b) By setting the determinant of $\overline{\overline{W}}(\omega, \mathbf{k})$ to zero, show that (4.41) can be satisfied by two different waves with the dispersion relations

$$k^2 = \omega^2 \mu_0 \varepsilon, \tag{4.42}$$

$$k_x^2 + k_y^2 + \frac{\varepsilon_z}{\varepsilon} k_z^2 = \omega^2 \mu_0 \varepsilon_z. \tag{4.43}$$

The wave governed by (4.43) is called the *extraordinary* electromagnetic wave since the relation between ω and $\mathbf{k}$ depends on the direction of $\mathbf{k}$. In contrast, the wave governed by (4.42) is isotropic and is called the *ordinary* electromagnetic wave.

(c) Assume the direction of $\mathbf{k}$ makes an angle θ with respect to the optic axis $\hat{z}$. Find the magnitudes of the phase velocities for these two waves for arbitrary θ.

(d) Using the results of parts (a) and (b), find the electric field polarizations of both waves.

Solution 4.3

The permittivity tensor of a uniaxial dielectric, with the optic axis along $\hat{z}$ is

$$\overline{\overline{\varepsilon}} = \begin{bmatrix} \varepsilon & 0 & 0 \\ 0 & \varepsilon & 0 \\ 0 & 0 & \varepsilon_z \end{bmatrix}.$$

We can also assume that the medium is nonmagnetic, i.e., $\overline{\overline{\mu}} = \mu_0 \overline{\overline{I}}$.

(a) From (4.34),

$$\left[\overline{\overline{k}} \cdot \overline{\overline{k}} + \omega^2 \mu_0 \overline{\overline{\varepsilon}} \right] \cdot \overline{E} = 0$$

$$\left[\overline{k}\,\overline{k} - k^2 \overline{\overline{I}} + \omega^2 \mu_0 \overline{\overline{\varepsilon}} \right] \cdot \overline{E} = 0$$

which we now rewrite as $\overline{\overline{W}} \left(\omega, \overline{k} \right) \cdot \overline{E} = 0$ where

$$\boxed{ \overline{\overline{W}} \left(\omega, \overline{k} \right) = \overline{k}\,\overline{k} - k^2 \overline{\overline{I}} + \omega^2 \mu_0 \overline{\overline{\varepsilon}}. } \tag{4.44}$$

(b) To evaluate the determinant of $\overline{\overline{W}}$, we first expand (4.44):

$$\overline{\overline{W}} = \begin{bmatrix} k_x^2 & k_x k_y & k_x k_z \\ k_y k_x & k_y^2 & k_y k_z \\ k_z k_x & k_z k_y & k_z^2 \end{bmatrix} - \begin{bmatrix} k^2 & 0 & 0 \\ 0 & k^2 & 0 \\ 0 & 0 & k^2 \end{bmatrix} + \begin{bmatrix} \omega^2 \mu_0 \varepsilon & 0 & 0 \\ 0 & \omega^2 \mu_0 \varepsilon & 0 \\ 0 & 0 & \omega^2 \mu_0 \varepsilon_z \end{bmatrix}$$

$$= \begin{bmatrix} \omega^2 \mu_0 \varepsilon - k_y^2 - k_z^2 & k_x k_y & k_x k_z \\ k_y k_x & \omega^2 \mu_0 \varepsilon - k_x^2 - k_z^2 & k_y k_z \\ k_z k_x & k_z k_y & \omega^2 \mu_0 \varepsilon_z - k_x^2 - k_y^2 \end{bmatrix}.$$

The determinant now becomes

$$
\begin{aligned}
|\overline{\overline{W}}| &= \left(\omega^2\mu_0\varepsilon - k_y^2 - k_z^2\right)\left(\omega^2\mu_0\varepsilon - k_x^2 - k_z^2\right)\left(\omega^2\mu_0\varepsilon_z - k_x^2 - k_y^2\right) + 2k_x^2k_y^2k_z^2 \\
&\quad - k_x^2k_z^2\left(\omega^2\mu_0\varepsilon - k_x^2 - k_z^2\right) - k_y^2k_z^2\left(\omega^2\mu_0\varepsilon - k_y^2 - k_z^2\right) \\
&\quad - k_x^2k_y^2\left(\omega^2\mu_0\varepsilon_z - k_x^2 - k_y^2\right) \\
&= \left(\omega^2\mu_0\varepsilon - k^2\right)\left(\omega^2\mu_0\varepsilon - k_x^2 - k_z^2\right)\left(\omega^2\mu_0\varepsilon_z - k_x^2 - k_y^2\right) \\
&\quad + k_x^2\left(\omega^2\mu_0\varepsilon - k_x^2 - k_z^2\right)\left(\omega^2\mu_0\varepsilon_z - k_x^2 - k_y^2\right) + 2k_x^2k_y^2k_z^2 \\
&\quad - k_x^2k_z^2\left(\omega^2\mu_0\varepsilon - k^2\right) - k_x^2k_y^2k_z^2 - k_y^2k_z^2\left(\omega^2\mu_0\varepsilon - k^2\right) \\
&\quad - k_x^2k_y^2k_z^2 - k_x^2k_y^2\left(\omega^2\mu_0\varepsilon_z - k_x^2 - k_y^2\right) \\
&= \left(\omega^2\mu_0\varepsilon - k^2\right)\left(\omega^2\mu_0\varepsilon - k_x^2 - k_z^2\right)\left(\omega^2\mu_0\varepsilon_z - k_x^2 - k_y^2\right) \\
&\quad + k_x^2\left(\omega^2\mu_0\varepsilon_z - k_x^2 - k_y^2\right)\left(\omega^2\mu_0\varepsilon - k^2\right) - k_z^2\left(\omega^2\mu_0\varepsilon - k^2\right)\left(k_x^2 + k_y^2\right) \\
&= \left(\omega^2\mu_0\varepsilon - k^2\right)\Big[\left(\omega^2\mu_0\varepsilon - k_x^2 - k_z^2\right)(\omega^2\mu_0\varepsilon_z - k_x^2 - k_y^2) \\
&\quad + k_x^2(\omega^2\mu_0\varepsilon_z - k_x^2 - k_y^2) - k_z^2(k_x^2 + k_y^2)\Big] \\
&= \left(\omega^2\mu_0\varepsilon - k^2\right)\Big[\left(\omega^2\mu_0\varepsilon_z - k_x^2 - k_y^2\right)(\omega^2\mu_0\varepsilon - k_z^2) - k_z^2(k_x^2 + k_y^2)\Big] \\
&= \left(\omega^2\mu_0\varepsilon - k^2\right)\Big[(\omega^2\mu_0)^2\varepsilon\varepsilon_z - \omega^2\mu_0\varepsilon(k_x^2 + k_y^2) - k_z^2\omega^2\mu_0\varepsilon_z \\
&\quad + k_z^2(k_x^2 + k_y^2) - k_z^2(k_x^2 + k_y^2)\Big] \\
|\overline{\overline{W}}| &= \left(\omega^2\mu_0\varepsilon - k^2\right)\Big[\omega^2\mu_0\varepsilon_z - (k_x^2 + k_y^2) - k_z^2\frac{\varepsilon_z}{\varepsilon}\Big]\omega^2\mu_0\varepsilon.
\end{aligned}
$$

Setting the determinant to zero yields two solutions:

$$
\boxed{
\begin{aligned}
\omega^2\mu_0\varepsilon &= k^2, &&\text{for the ordinary wave,} \\
\omega^2\mu_0\varepsilon_z &= k_x^2 + k_y^2 + k_z^2\frac{\varepsilon_z}{\varepsilon}, &&\text{for the extraordinary wave.}
\end{aligned}
}
$$

(c) The phase velocities for ordinary and extraordinary waves show different behaviors.

- The dispersion relation of the ordinary wave has only a dependence on the magnitude of the wave vector. Hence, it has the same phase velocity in all directions:

$$
\boxed{v_p = \frac{\omega}{k} = \frac{1}{\sqrt{\mu_0\varepsilon}}.}
$$

- In contrast, the dispersion relation of the extraordinary wave depends on k_z differently from the components of k transverse to $\hat{z}$. Hence, the dispersion relation changes with the angle of propagation, as does the phase velocity:

$$\omega^2 \mu_0 \varepsilon_z = k^2 \sin^2 \theta + k^2 \cos^2 \theta \frac{\varepsilon_z}{\varepsilon}$$

$$= k^2 \left[\sin^2 \theta + \frac{\varepsilon_z}{\varepsilon} \cos^2 \theta \right]$$

$$\therefore \ v_\mathrm{p} = \frac{\omega}{k} = \frac{1}{\sqrt{\mu_0 \varepsilon_z}} \left[\sin^2 \theta + \frac{\varepsilon_z}{\varepsilon} \cos^2 \theta \right]^{\frac{1}{2}}$$

(d) Let us consider the ordinary and extraordinary waves separately.

- For the *ordinary wave*

$$\overline{\overline{W}} = \begin{bmatrix} k_x^2 & k_x k_y & k_x k_z \\ k_y k_x & k_y^2 & k_y k_z \cdot \\ k_z k_x & k_z k_y & k_z^2 + \omega^2 \mu_0 (\varepsilon_z - \varepsilon) \end{bmatrix}$$

choose $\overline{k} = k \cos \theta \, \hat{z} + k \sin \theta \, \hat{x}$ such that

$$\overline{\overline{W}} \cdot \overline{E} = \begin{bmatrix} k^2 \sin^2 \theta & 0 & k^2 \sin \theta \cos \theta \\ 0 & 0 & 0 \\ k^2 \sin \theta \cos \theta & 0 & k^2 \cos^2 \theta + \omega^2 \mu_0 (\varepsilon_z - \varepsilon) \end{bmatrix} \begin{bmatrix} E_x \\ E_y \\ E_z \end{bmatrix} = 0.$$

Therefore,

$$\begin{bmatrix} \sin^2 \theta & 0 & \sin \theta \cos \theta \\ 0 & 0 & 0 \\ \sin \theta \cos \theta & 0 & \cos^2 \theta + \frac{\varepsilon_z - \varepsilon}{\varepsilon} \end{bmatrix} \begin{bmatrix} E_x \\ E_y \\ E_z \end{bmatrix} = 0.$$

This reduces to a 2 × 2 matrix times a field column vector containing only E_x and E_z. The determinant of this 2 × 2 matrix is

$$\sin^2 \theta \left(\cos^2 \theta + \frac{\varepsilon_z - \varepsilon}{\varepsilon} \right) - \sin^2 \theta \cos^2 \theta = 0$$

with the two possibilities,

$$\sin \theta = 0 \quad \text{or} \quad \left[\frac{\varepsilon_z - \varepsilon}{\varepsilon} \right] = 0.$$

Fig. 4.1 Plane wave
propagating along the
direction $\bar{k}$ with the electric
field along $\hat{y}$

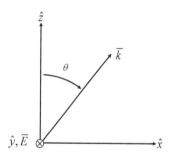

There is no general solution to this equation for $\varepsilon_z \neq \varepsilon$ and θ arbitrary. Thus, both E_x and E_z must be zero. There are no restrictions on E_y, however, so

$$\boxed{\bar{E} = E_0\,\hat{y}.}$$

In general, $\bar{E} \propto \hat{z} \times \bar{k}$ for the ordinary wave, as shown in Fig. 4.1. The reason why the permittivity along the $\hat{z}$ direction (ε_z) does not enter the dispersion relation is because there is no component of $\bar{E}$ along $\hat{z}$.

- For the *extraordinary* wave, as before choose $\bar{k} = k\left(\hat{z}\cos\theta + \hat{x}\sin\theta\right)$

$$\begin{bmatrix} \omega^2\mu_0\varepsilon - k^2\cos\theta & 0 & k^2\cos\theta\sin\theta \\ 0 & \omega^2\mu_0\varepsilon - k^2 & 0 \\ k^2\cos\theta\sin\theta & 0 & \omega^2\mu_0\varepsilon_z - k^2\sin^2\theta \end{bmatrix} \begin{bmatrix} E_x \\ E_y \\ E_z \end{bmatrix} = 0$$

Since $\omega^2\mu_0\varepsilon - k^2 \neq 0$, we must have $E_y = 0$ and the remaining two field components must satisfy

$$\begin{bmatrix} \dfrac{\varepsilon}{\varepsilon_z}k^2\sin^2\theta & k^2\sin\theta\cos\theta \\ k^2\sin\theta\cos\theta & k^2\dfrac{\varepsilon}{\varepsilon_z}\cos^2\theta \end{bmatrix} \begin{bmatrix} E_x \\ E_z \end{bmatrix} = 0,$$

where we have used the extraordinary wave dispersion relation from part (b). From the first row of this matrix equation we have

$$E_x \frac{\varepsilon}{\varepsilon_z}\sin^2\theta = -E_z\cos\theta\sin\theta$$

$$\text{or,} \quad \frac{E_z}{E_x} = -\frac{\varepsilon}{\varepsilon_z}\frac{\sin\theta}{\cos\theta}.$$

Similarly, the second row gives

$$E_x\cos\theta\sin\theta = -E_z\cos^2\theta\frac{\varepsilon_z}{\varepsilon}$$

$$\frac{E_z}{E_x} = -\frac{\varepsilon}{\varepsilon_z}\frac{\sin\theta}{\cos\theta}$$

Fig. 4.2 Direction of $\overline{D}$ for the extraordinary wave traveling at the angle θ with the z axis

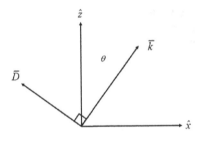

$$\therefore \ \overline{E} = \frac{E_0}{\sqrt{(\varepsilon_z/\varepsilon)^2 \cos^2 \theta + \sin^2 \theta}} \begin{bmatrix} -\dfrac{\varepsilon_z}{\varepsilon} \cos \theta \\ 0 \\ \sin \theta \end{bmatrix}.$$

In general, $\overline{D} \propto \overline{k} \times (\hat{z} \times \overline{k})$, and $\overline{D} = \overline{\overline{\varepsilon}} \cdot \overline{E}$, as shown in Fig. 4.2.

Problem 4.4 Consider wave propagation along the direction of an applied magnetic field in a ferrite, as described in [5, Sect. 4.9.1]. If the field and propagation are along the $\hat{z}$ direction, the permeability is of the form (4.12). Assume the permittivity is isotropic. The normal modes are found to be circularly polarized, with the time-varying component of the magnetic field of the form $h_y/h_x = \pm i$. The dispersion relations for the modes are given by

$$k_+^2 = k_0^2 \left[\frac{\omega_0 + \omega_M - \omega}{\omega_0 - \omega} \right], \tag{4.45}$$

$$k_-^2 = k_0^2 \left[\frac{\omega_0 + \omega_M + \omega}{\omega_0 + \omega} \right]. \tag{4.46}$$

Plot the dispersion relations for the $\pm$ modes with ω/ω_M along the vertical axis, and $\mathrm{Re}\{k_\pm/k_M\}$ along the horizontal axis, where $k_M = \omega_M \sqrt{\mu_0 \varepsilon}$ and $k_0 = \omega \sqrt{\mu_0 \varepsilon}$. For the plot, take $\omega_0/\omega_M = 1$. (Note that the waves do not propagate for k imaginary.)

Solution 4.4

It is convenient to rewrite (4.45) and (4.46) in terms of the normalized quantities $k_\pm/k_M$, ω/ω_M, and ω_0/ω_M:

$$\frac{k_+^2}{k_M} = \frac{\omega}{\omega_M} \left[\frac{\omega_0/\omega_M + 1 - \omega/\omega_M}{\omega_0/\omega_M - \omega/\omega_M} \right], \tag{4.47}$$

$$\frac{k_-^2}{k_M} = \frac{\omega}{\omega_M} \left[\frac{\omega_0/\omega_M + 1 + \omega/\omega_M}{\omega_0/\omega_M + \omega/\omega_M} \right]. \tag{4.48}$$

A Python program to plot the dispersion relations is given below, with the resulting plot shown in Fig. 4.3. The horizontal solid line at $\omega/\omega_M = 1$ in the plot is a graphing artifact from the singularity in the denominator of (4.47), but also can be taken to visually represent the frequency asymptote as $k_+/k_M \to \infty$.

```python
from matplotlib import pyplot as plt
import numpy as np
from cmath import sqrt

# calculate the dispersion relation of plane waves
# propagating along the field axis of a magnetized
# magnetic insulator, e.g., a ferrite

# define vectorized functions

def mysqrt(x):
  return sqrt(x)

def myreal(x):
  return x.real

mysqrtv = np.vectorize(mysqrt)
myrealv = np.vectorize(myreal)

# define parameters: Z = \omega_0/\omega_M,
# OM = \omega/\omega_M
Z = 1
OMmax = 5

# number of points N
N = 512

# create normalized frequency array

OM = np.arange(1e-3,OMmax,OMmax/(N-1))

# calculate dispersion relations

kp = OM*mysqrtv((Z+1-OM)/(Z-OM))
km = OM*mysqrtv((Z+1+OM)/(Z+OM))

# plot
fig, ax=plt.subplots()
ax.plot(kp,OM,label='$+$')
ax.plot(km,OM,linestyle='--',label='$-$')
ax.set(xlabel='normalized wave number ($k_{\pm}/k_M$)',
       ylabel='normalized frequency ($\omega/\omega_M$)')
ax.grid()
ax.set_xlim(0,10)
ax.set_ylim(0,5)
ax.tick_params(axis='both', direction='in')
ax.legend(loc='upper right')
fig1 = plt.gcf()
plt.show()
```

Fig. 4.3 Dispersion relations for right- and left-circularly polarized electromagnetic waves propagating in a magnetized ferrite, with the propagation direction parallel to the applied magnetic field

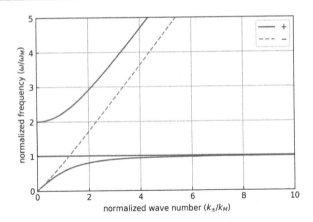

normalized wave number ($k_\pm/k_M$)

Problem 4.5 Consider wave propagation along the direction of an applied magnetic field in a ferrite, as explained in [5, Sect. 4.9.1]. If propagation occurs along the $\hat{\mathbf{z}}$ direction, assume the total electric field at $z = 0$ is linearly polarized along the $\hat{\mathbf{x}}$ direction.

(a) Show that the linearly polarized field can be expressed as the sum of left and right circularly polarized waves propagating with wave numbers k_- and k_+, respectively.
(b) Show that after propagating a distance d, the linear polarization of the total field will be rotated through the angle (Fig. 4.4)

$$\theta_F = (k_+ - k_-)\frac{d}{2}. \tag{4.49}$$

This phenomenon is called *Faraday rotation*.
(c) Show that the polarization continues to rotate in the same direction if the propagation direction reverses.
(d) Using the dispersion relations given in Problem 4.4, show that the direction of rotation for $\omega > \omega_0 + \omega_M$ is opposite to that for $\omega < \omega_0$.

Fig. 4.4 Faraday rotation angle for a wave initially polarized along the x axis, with the propagation direction and an applied magnetic bias field along the z axis. (Reproduced from [5, Sect. 4.9.1] with permission)

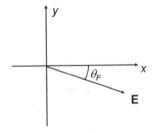

Solution 4.5
Linearly polarized electric fields can be written as a combination of right- and left-circularly polarized fields:

$$\overline{E} = \frac{E_0}{2} \underbrace{\left[\hat{x} + i\hat{y}\right] e^{ik_+ z}}_{\text{RHCP}} + \frac{E_0}{2} \underbrace{\left[\hat{x} - i\hat{y}\right] e^{ik_- z}}_{\text{LHCP}} \tag{4.50}$$

(a) At $z = 0$, (4.50) becomes

$$\overline{E}\,(z = 0) = \frac{E_0}{2} \left[\hat{x} + i\hat{y} + \hat{x} - i\hat{y}\right]$$
$$= E_0 \hat{x}$$

thus establishing that linearly polarized fields can be represented using a basis set defined by the two orthogonal circular polarizations.
(b) To calculate the electric field after propagation through a distance d, we set $z = d$ in:

$$\overline{E}\,(z = d) = \frac{E_0}{2} \left[\hat{x} + i\hat{y}\right] e^{ik_+ d} + \frac{E_0}{2} \left[\hat{x} - i\hat{y}\right] e^{ik_- d}$$

$$= E_0 \left(\frac{e^{ik_+ d} + e^{ik_- d}}{2}\right) \hat{x} + i E_0 \left(\frac{e^{ik_+ d} - e^{ik_- d}}{2}\right) \hat{y}$$

$$= \begin{aligned} &E_0 e^{i(k_+ + k_-)\frac{d}{2}} \left(\dfrac{e^{i(k_+ - k_-)\frac{d}{2}} + e^{-i(k_+ - k_-)\frac{d}{2}}}{2}\right) \hat{x} \\ &- E_0 e^{i(k_+ + k_-)\frac{d}{2}} \left(\dfrac{e^{i(k_+ - k_-)\frac{d}{2}} - e^{-i(k_+ - k_-)\frac{d}{2}}}{2i}\right) \hat{y} \end{aligned}$$

$$= E_0 e^{i(k_+ + k_-)\frac{d}{2}} \left[\hat{x} \cos\left[(k_+ - k_-)\frac{d}{2}\right] - \hat{y} \sin\left[(k_+ - k_-)\frac{d}{2}\right]\right]$$

Thus the resultant $\overline{E}$ field is still linear but makes an angle θ_F with respect to the $\hat{x}$ axis:

$$\tan \theta_F = \tan (k_+ - k_-)\frac{d}{2}$$

Therefore,

$$\boxed{\theta_F = (k_+ - k_-)\frac{d}{2}.}$$

(c) To see how the field would have rotated if we propagated along $-\hat{z}$, we make the substitution:

$$k_\pm \to -k_\pm, \text{ and } d \to -d:$$

Thus,

$$\bar{E}\,(z=-d) = E_0 e^{+i(k_+ + k)\frac{d}{2}} \left[\hat{x} \cos\left[(k_+ - k_-)\frac{d}{2}\right] - \hat{y} \sin\left[(k_+ - k_-)\frac{d}{2}\right] \right]$$

There is no change! Thus the polarization rotates in the same direction; it is not possible to undo the rotation by reflecting the wave back through the medium.

(d) Clearly the sign of θ_F depends on the sign of $k_+ - k_-$. From (4.45) and (4.46) we note that the sign of $(|\,k_+\,| - |\,k_-\,|)$ is the same as the sign of $\left(k_+^2 - k_-^2\right)$. Thus:

$$
\begin{aligned}
\frac{k_+^2 - k_-^2}{k_0^2} &= \frac{\omega_0 + \omega_M - \omega}{\omega_0 - \omega} - \frac{\omega_0 + \omega_M + \omega}{\omega_0 + \omega} \\
&= \frac{(\omega_0 - \omega)(\omega_0 + \omega) + \omega_M(\omega_0 + \omega)}{\omega_0^2 - \omega^2} \\
&\quad - \frac{(\omega_0 + \omega)(\omega_0 - \omega) - \omega_M(\omega_0 - \omega)}{\omega_0^2 - \omega^2} \\
&= \frac{2\omega\omega_M}{\omega_0^2 - \omega^2}.
\end{aligned}
$$

Consider the two cases

- $\omega < \omega_0$

$$\frac{k_+^2 - k_-^2}{k_0^2} > 0, \quad \text{or} \quad k_+ > k_- \Rightarrow \theta_F > 0$$

- $\omega > \omega_0 + \omega_M$

$$\frac{k_+^2 - k_-^2}{k_0^2} < 0 \quad \text{or} \quad k_+ < k_- \Rightarrow \theta_F < 0$$

Note that k_+ doesn't propagate for $\omega_0 < \omega < \omega_0 + \omega_M$ (See Fig. 4.3.)

Problem 4.6 Consider wave propagation perpendicular to an applied magnetic field in a ferrite, as explained in [5, Sect. 4.9.2]. Specifically, let the magnetic bias field be along $\hat{z}$ and propagation along $\hat{x}$. The equation for the x and y magnetic field components of the wave associated with the dispersion relation

$$k_a^2 = k_0^2 = \omega^2 \mu_0 \varepsilon \tag{4.51}$$

is obtained by substituting $k = k_a$ into (4.33), which in this case becomes

$$
\begin{bmatrix}
k_0^2(1+\chi) - k^2 & -ik_0^2\kappa & 0 \\
+ik_0^2\kappa & k_0^2(1+\chi) & 0 \\
0 & 0 & k_0^2 - k^2
\end{bmatrix}
\begin{bmatrix}
h_x \\ h_y \\ h_z
\end{bmatrix} = 0. \tag{4.52}
$$

(a) Show that the x and y components of the magnetic fields for the k_a wave must satisfy

$$\begin{bmatrix} \chi & -i\kappa \\ i\kappa & 1+\chi \end{bmatrix} \begin{bmatrix} h_x \\ h_y \end{bmatrix} = 0. \tag{4.53}$$

(b) Using the definitions for χ and κ,

$$\chi = \frac{\omega_0\omega_M}{\omega_0^2 - \omega^2}, \quad \kappa = \frac{\omega\omega_M}{\omega_0^2 - \omega^2},$$

show that the only solution for finite ω is $h_x = h_y = 0$, and consequently the only nonzero component of h for this mode is h_z. (Hint: If the elements of $\overline{\overline{\chi}}$ are singular, $\mathbf{h} = 0$ for any finite $\mathbf{m}$ since $\mathbf{m} = \overline{\overline{\chi}} \cdot \mathbf{h}$.)

Solution 4.6

Consider plane wave propagation perpendicular to the applied field in a ferrite sample.

(a) Substituting (4.51) into (4.52) gives

$$\begin{bmatrix} k_0^2\chi & -ik_0^2\kappa & 0 \\ ik_0^2\kappa & k_0^2(1+\chi) & 0 \\ 0 & 0 & 0 \end{bmatrix} \begin{bmatrix} h_x \\ h_y \\ h_z \end{bmatrix} = 0.$$

Dividing out the k_0^2 term yields

$$\boxed{\begin{bmatrix} \chi & -i\kappa \\ i\kappa & (1+\chi) \end{bmatrix} \begin{bmatrix} h_x \\ h_y \end{bmatrix} = 0.}$$

(b) For a nontrivial solution, set $\det\left[\overline{\overline{\chi}}\right] = 0$:

$$\chi(1+\chi) = \kappa^2$$

or, $$\frac{\omega_0\omega_M(\omega_0^2 - \omega^2 + \omega_0\omega_M)}{\left(\omega_0^2 - \omega^2\right)^2} = \frac{\omega^2\omega_M^2}{\left(\omega_0^2 - \omega^2\right)^2}$$

If we set the numerators equal,

$$\omega_0\omega_M(\omega_0^2 - \omega^2 + \omega_0\omega_M) = \omega^2\omega_M^2$$
$$\omega_0^2\omega_M(\omega_0 + \omega_M) = \omega^2(\omega_M^2 + \omega_0\omega_M)$$
$$\omega_0^2 = \omega^2$$

But at this frequency χ and κ are singular; if $\overline{m} = \overline{\overline{\chi}} \cdot \overline{h}$, then $\overline{h} = [h_x, \, h_y]^T = 0$ for any finite $\overline{m}$ if the elements of $\overline{\overline{\chi}}$ are singular. We conclude that for this mode the magnetic field $\overline{h}$ is polarized along $\hat{z}$.

References

1. F.T. Ulaby, U. Ravaioli, *Fundamentals of Applied Electromagnetics*, 7th edn. (Pearson, Essex, 2015)
2. H.C. Chen, *Theory of Electromagnetic Waves: A Coordinate-Free Approach* (McGraw-Hill, New York, 1983)
3. H.A. Haus, *Waves and Fields in Optoelectronics* (Prentice-Hall, Englewood Cliffs, 1984)
4. J.A. Kong, *Electromagnetic Wave Theory* (EMW Publishing, Cambridge, 2005)
5. D.D. Stancil, A. Prabhakar, *Spin Waves: Theory and Applications* (Springer, New York, 2009)

Magnetostatic Modes

5

We have previously described the theory of electromagnetic wave propagation in isotropic and anisotropic media. In this chapter, we introduce the *magnetostatic approximation* to Maxwell's equations, and use its solutions to analyze the different modes of propagation in a variety of geometries.[1]

The magnetostatic approximation, the Polder susceptibility tensor, and the spin wave manifold are first introduced and then explored in Problem 5.1. The use of demagnetizing fields to find the magnetostatic uniform precession modes of various geometries is the topic of Problems 5.2 and 5.3. Propagating modes in the magnetostatic approximation are referred to as *dipolar spin waves* when the dynamical contribution of the exchange interaction is negligible. The geometries and properties of forward, backward, and surface dipolar spin wave modes in thin films are the subjects of Problems 5.4–5.7. These modes are then modified to include the effects of exchange interactions, and we learn about the effects of surface pinning through Problem 5.8.

5.1 The Magnetostatic Approximation

Consider a uniform plane wave of the form $e^{i\mathbf{k}\cdot\mathbf{r}}$, propagating in an arbitrary direction in a magnetized ferrite. We can now write Maxwell's equations (4.3) as

$$\mathbf{k} \times \mathbf{h} = -\omega\varepsilon\mathbf{e}, \tag{5.1}$$

$$\mathbf{k} \times \mathbf{e} = \omega\mu_0(\mathbf{h} + \mathbf{m}). \tag{5.2}$$

---

[1] Further references on these topics include Lax and Button [1], Soohoo [2], and Sodha and Srivastava [3].

© The Author(s), under exclusive license to Springer Nature Switzerland AG 2021
D. D. Stancil and A. Prabhakar, *Spin Waves*,
https://doi.org/10.1007/978-3-030-68582-9_5

Crossing $\mathbf{k}$ into both sides of (5.1), and substituting (5.2) for $\mathbf{k} \times \mathbf{e}$, we get

$$\mathbf{kk} \cdot \mathbf{h} - k^2 \mathbf{h} = -\omega^2 \mu_0 \varepsilon (\mathbf{h} + \mathbf{m}), \qquad (5.3)$$

where we have also used the identity

$$\mathbf{k} \times \mathbf{k} \times \mathbf{I} = \overline{\mathbf{k}} \cdot \overline{\mathbf{k}} = \mathbf{kk} - k^2 \overline{\mathbf{I}}. \qquad (5.4)$$

Since

$$\mathbf{k} \cdot \mathbf{b} = \mu_0 \mathbf{k} \cdot (\mathbf{h} + \mathbf{m}) = 0,$$

we have that

$$\mathbf{k} \cdot \mathbf{h} = -\mathbf{k} \cdot \mathbf{m}. \qquad (5.5)$$

Substituting this expression for $\mathbf{k} \cdot \mathbf{h}$ into (5.3) and solving for $\mathbf{h}$ yields

$$\mathbf{h} = \frac{k_0^2 \mathbf{m} - \mathbf{kk} \cdot \mathbf{m}}{k^2 - k_0^2}, \qquad (5.6)$$

where $k_0^2 = \omega^2 \mu_0 \varepsilon$.

To find $\mathbf{e}$, we cross $\mathbf{k}$ into (5.2), use $\mathbf{k} \cdot \mathbf{e} = 0$ and substitute (5.1) for $\mathbf{k} \times \mathbf{h}$. This yields

$$-k^2 \mathbf{e} = -\omega^2 \mu_0 \varepsilon \, \mathbf{e} + \omega \mu_0 \, \mathbf{k} \times \mathbf{m}, \qquad (5.7)$$

Solving for $\mathbf{e}$ gives

$$\mathbf{e} = \frac{\omega \mu_0 \, \mathbf{k} \times \mathbf{m}}{k_0^2 - k^2}. \qquad (5.8)$$

Recognizing that $\nabla \times \mathbf{h} = -\omega \varepsilon \, \mathbf{e}$ enables us to write

$$\nabla \times \mathbf{h} = -\frac{k_0^2 \, \mathbf{k} \times \mathbf{m}}{k_0^2 - k^2}. \qquad (5.9)$$

Let us examine (5.6), (5.8), and (5.9) in the limit $|\mathbf{k}| \gg |\mathbf{k_0}|$. Note that both the numerator and denominator of (5.6) contain terms that are quadratic in $\mathbf{k}$. As a result, provided that $\mathbf{k} \cdot \mathbf{m} \neq 0$, $\mathbf{h}$ remains finite as $|\mathbf{k}|$ becomes large. On the other hand, (5.8) and (5.9) go as $1/k$, and therefore vanish in the limit of large $|\mathbf{k}|$. We conclude that even though microwave time variations are involved, the fields satisfy the magnetostatic equations

$$\nabla \times \mathbf{h} = 0, \qquad (5.10a)$$
$$\nabla \cdot \mathbf{b} = 0. \qquad (5.10b)$$

Once these equations are solved for **b** and **h**, we can obtain a first-order approximation to **e** from

$$\nabla \times \mathbf{e} = i\omega\,\mathbf{b}. \tag{5.11}$$

Equations (5.10a), (5.10b), and (5.11) comprise the magnetoquasistatic approximation to Maxwell's equations.[2] Waves that are described by these equations are commonly referred to as *magnetostatic waves*, although magnetoquasistatic waves would be more precise.

Since $\nabla \times \mathbf{h} = 0$, we can introduce the *magnetostatic scalar potential* ψ,

$$\mathbf{h} = -\nabla\psi, \tag{5.12}$$

and combine it with the quasistatic equations (5.10a) and (5.10b) to obtain *Walker's equation*

$$(1+\chi)\left[\frac{\partial^2\psi}{\partial x^2} + \frac{\partial^2\psi}{\partial y^2}\right] + \frac{\partial^2\psi}{\partial z^2} = 0. \tag{5.13}$$

Inherent in this derivation is the assumption that χ and κ, in the permeability tensor

$$\overline{\overline{\mu}} = \mu_0 \begin{bmatrix} 1+\chi & -i\kappa & 0 \\ i\kappa & 1+\chi & 0 \\ 0 & 0 & 1 \end{bmatrix}, \tag{5.14}$$

are independent of position. The solutions to Walker's equations are commonly referred to as *magnetostatic modes*.

If we consider a uniform plane wave, $\psi \propto e^{i\mathbf{k}\cdot\mathbf{r}}$, propagating in an infinite medium, (5.13) becomes

$$(1+\chi)(k_x^2 + k_y^2) + k_z^2 = 0. \tag{5.15}$$

Assuming the direction of propagation makes the angle θ with the $\hat{\mathbf{z}}$ axis (also the direction of the DC bias field), the components of **k** can be written

$$k_x^2 + k_y^2 = k^2 \sin^2\theta, \tag{5.16}$$

$$k_z^2 = k^2 \cos^2\theta. \tag{5.17}$$

Substituting these equations into (5.15) and simplifying yields

$$\chi \sin^2\theta = -1. \tag{5.18}$$

Expressing χ in terms of frequency using (3.63a) leads to the *dispersion relation* (cf. Problem 3.6)

$$\omega = [\omega_0(\omega_0 + \omega_M \sin^2\theta)]^{1/2}. \tag{5.19}$$

---

[2]Examination of (5.6), (5.8), and (5.9) shows that the magnetoquasistatic approximation is also valid in the limit $|\mathbf{k}| \ll |\mathbf{k}_0|$.

The spin wave *manifold* refers to all the propagating waves in the frequency range

$$\omega_0 \leq \omega \leq \sqrt{\omega_0(\omega_0 + \omega_M)}. \tag{5.20}$$

Problem 5.1 Consider plane waves propagating in a magnetized ferrite.

(a) Show that the magnetic field in the $|\mathbf{k}| \gg |\mathbf{k}_0|$ magnetostatic limit is given by

$$\mathbf{h} = -\frac{\mathbf{k}\mathbf{k} \cdot \mathbf{m}}{k^2}. \tag{5.21}$$

(b) Using the constitutive relation between $\mathbf{m}$ and $\mathbf{h}$ along with the results of part (a), show that $\mathbf{h}$ must satisfy

$$(k^2\overline{\mathbf{I}} + \mathbf{k}\mathbf{k} \cdot \overline{\chi}) \cdot \mathbf{h} = 0, \tag{5.22}$$

where $\overline{\chi}$ is the Polder susceptibility tensor

$$\overline{\chi} = \begin{bmatrix} \chi & -i\kappa \\ i\kappa & \chi \end{bmatrix}.$$

(c) Assume $\mathbf{k} = k(\hat{\mathbf{y}} \sin\theta + \hat{\mathbf{z}}\cos\theta)$, where θ is the angle between $\mathbf{k}$ and the applied field direction (taken to be $\hat{\mathbf{z}}$). Substitute this into the result of part (b) and show that the existence of a nontrivial solution requires[3]

$$\omega^2 = \omega_0(\omega_0 + \omega_M \sin^2\theta). \tag{5.23}$$

The limiting cases of $\theta = 0$ and $\theta = \pi/2$ define the frequency band over which magnetostatic plane wave type excitations can exist.

Solution 5.1

We are considering plane wave propagation in the magnetostatic limit.

(a) For $|\overline{k}| \gg |\overline{k}_0|$,

$$\boxed{\overline{h} = \frac{k_0^2\overline{m} - \overline{k}\,\overline{k} \cdot \overline{m}}{k^2 - k_0^2} \approx \frac{-\overline{k}\,\overline{k} \cdot \overline{m}}{k^2}.}$$

---

[3] This expression gives the correct frequencies for the $k \to \infty$ asymptotes for the limiting cases of $\theta = 0$, $\pi/2$ discussed in Stancil and Prabhakar [4, Sect. 4.9]. The present result shows that the frequency of the asymptote lies between these two limits for intermediate angles.

(b) The constitutive relation for a magnetic material is $\overline{m} = \overline{\overline{\chi}} \cdot \overline{h}$. Thus,

$$\overline{h} = -\frac{\overline{k}\,\overline{k}}{k^2} \cdot \overline{\overline{\chi}} \cdot \overline{h} \quad \Rightarrow \quad k^2\overline{h} + \overline{k}\,\overline{k} \cdot \overline{\overline{\chi}} \cdot \overline{h} = 0.$$

Factor out $\overline{h}$ to get

$$\boxed{\left[k^2\overline{\overline{I}} + \overline{k}\,\overline{k} \cdot \overline{\overline{\chi}}\right] \cdot \overline{h} = 0.}$$

(c) Let, $\overline{k} = k\left(\hat{y}\sin\theta + \hat{z}\cos\theta\right)$, such that the coefficient matrix becomes

$$C = \begin{bmatrix} k^2 & 0 & 0 \\ 0 & k^2 & 0 \\ 0 & 0 & k^2 \end{bmatrix} + \begin{bmatrix} 0 & 0 & 0 \\ 0 & k_y^2 & k_y k_z \\ 0 & k_z k_y & k_z^2 \end{bmatrix} \begin{bmatrix} \chi & -i\kappa & 0 \\ i\kappa & \chi & 0 \\ 0 & 0 & 0 \end{bmatrix}$$

$$= \begin{bmatrix} k^2 & 0 & 0 \\ 0 & k^2 & 0 \\ 0 & 0 & k^2 \end{bmatrix} + \begin{bmatrix} 0 & 0 & 0 \\ i\kappa k^2\sin^2\theta & \chi k^2\sin^2\theta & 0 \\ i\kappa k^2\sin\theta\cos\theta & \chi k^2\sin\theta\cos\theta & 0 \end{bmatrix}$$

Dividing out k^2, we have

$$\begin{bmatrix} 1 & 0 & 0 \\ i\kappa\sin^2\theta & 1 + \chi\sin^2\theta & 0 \\ i\kappa\sin\theta\cos\theta & \chi\sin\theta\cos\theta & 1 \end{bmatrix} \begin{bmatrix} h_x \\ h_y \\ h_z \end{bmatrix} = 0$$

If we set the determinant to zero, we get the condition $1 + \chi\sin^2\theta = 0$. Substituting for χ yields

$$\boxed{\omega^2 = \omega_0(\omega_0 + \omega_M\sin^2\theta).}$$

5.2 Uniform Precession Modes

If the magnetization is uniform in the sample, then the spin dynamics is dominated by the applied field and a *demagnetizing field* resulting from effective charges where the magnetization vector terminates at a surface. The strength of this demagnetizing field depends on the geometry, and is described by a demagnetizing factor (or in general a demagnetizing tensor). In this section, we consider the form of the demagnetizing field for a few important geometries.

For samples that are ellipsoidal or limiting cases of ellipsoids, the demagnetizing field from the effective magnetic surface charges arising from a uniform magnetization is also uniform. In such cases, the demagnetizing field from the magnetic surface charges of the sample can be written

$$\mathbf{H_d} = -\overline{\mathbf{N}}_d \cdot \mathbf{M}, \tag{5.24}$$

where $\overline{\mathbf{N}}_d$ is referred to as the demagnetizing tensor. This tensor has the property that the trace is always equal to 1.

As an example, consider a uniformly magnetized sphere. Owing to the symmetry, the demagnetizing tensor is given by

$$\overline{\mathbf{N}}_d = \begin{bmatrix} 1/3 & 0 & 0 \\ 0 & 1/3 & 0 \\ 0 & 0 & 1/3 \end{bmatrix}, \tag{5.25}$$

and the total internal field is given by

$$H_0 = H_{DC} - \frac{1}{3} M_S \tag{5.26}$$

regardless of the direction of $\mathbf{H_{DC}}$, provided only that H_0 is positive.[4]

Now let us consider small time-varying components of magnetization about the static equilibrium direction. Taking $\mathbf{H_{DC}}$ along $\hat{z}$ and assuming the small-signal response is uniform, we must simultaneously satisfy

$$\mathbf{h} = \overline{\chi}^{-1} \cdot \mathbf{m} = \frac{1}{\omega_M} \begin{bmatrix} \omega_0 & i\omega \\ -i\omega & \omega_0 \end{bmatrix} \cdot \begin{bmatrix} m_x \\ m_y \end{bmatrix} \tag{5.27}$$

and

$$\mathbf{h} = -\overline{\mathbf{N}}_d \cdot \mathbf{m}. \tag{5.28}$$

In general, the resonant frequency of the uniform precession mode is obtained from setting (5.27) equal to (5.28) and requiring the determinant of the coefficient matrix to vanish:

$$\det\left[\overline{\mathbf{N}}_d + \overline{\chi}^{-1}\right] = 0. \tag{5.29}$$

In the case of a sphere we have

$$\begin{vmatrix} \omega_0 + \omega_M/3 & i\omega \\ -i\omega & \omega_0 + \omega_M/3 \end{vmatrix} = 0, \tag{5.30}$$

leading to

$$\omega = \omega_0 + \omega_M/3. \tag{5.31}$$

---

[4]This is necessary to ensure that the magnetization of the sphere is saturated.

Using (5.26) gives

$$\omega = -\gamma\mu_0 \left[H_{DC} - \frac{1}{3}M_S + \frac{1}{3}M_S \right]$$

$$= -\gamma\mu_0 H_{DC}.$$

(5.32)

Thus the effects of the RF and DC demagnetizing fields cancel, and the resonant frequency depends only on the strength of the applied field. The fact that the frequency does not depend on material properties is attractive for device applications.

As a second example, consider a magnetized thin film. Since a thin film can be viewed as the limiting case of an oblate spheroid, the internal demagnetizing field is also uniform when the magnetization is uniform, and the demagnetizing tensor can again be used to find the demagnetizing field. For example, when $\mathbf{H}_{DC}$ is along $\hat{z}$, the film is in the $x - z$ plane, and the small-signal magnetization response is uniform, only the y component of the small-signal magnetization is reduced by the demagnetizing factor, and we must have

$$\mathbf{h} = -\overline{\mathbf{N}}_d \cdot \mathbf{m} = - \begin{bmatrix} 0 & 0 \\ 0 & 1 \end{bmatrix} \begin{bmatrix} m_x \\ m_y \end{bmatrix}.$$

(5.33)

Setting (5.27) equal to (5.33) and requiring the determinant of the resulting matrix to vanish leads to

$$\omega = [\omega_0(\omega_0 + \omega_M)]^{1/2}.$$

(5.34)

This is the resonant frequency of the uniform precession mode of a thin film with in-plane magnetization.

Problem 5.2 Calculate the uniform precession frequency for a YIG film in the applied field $B_{DC} = 0.5\,\mathrm{T}$ if the field is

(a) parallel to the film, and
(b) perpendicular to the film.

For YIG, $M_S = 140\,\mathrm{kA/m}$, and $|\gamma|/2\pi = 28\,\mathrm{GHz/T}$.

Solution 5.2

(a) The uniform precession frequency when the applied field is parallel to the film is

$$\omega_\| = \sqrt{\omega_0(\omega_0 + \omega_M)}$$

$$f_\| = \left| \frac{\gamma}{2\pi} \right| \sqrt{B_{DC}(B_{DC} + \mu_0 M_S)}$$

$$= 28\sqrt{0.5(0.5 + 0.176)}$$

$$\boxed{f_\| = 16.28\,\mathrm{GHz}}$$

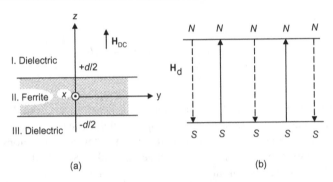

Fig. 5.1 **a** Geometry for the uniform precession mode analysis of a normally magnetized ferrite film. **b** The demagnetizing field generated by effective magnetic surface charges (Reproduced from [4], with permission)

(b) In the case where the applied field is perpendicular to the film plane, the surface charges generate an effective demagnetizing field $\overline{H}_{DC}$ that is equal and opposite to $\hat{z}M_S$, as shown in Fig. 5.1. Consequently, the demagnetization field must be subtracted from the applied field, and we have a lower frequency for the uniform precession:

$$f_\perp = \left| \frac{\gamma}{2\pi} \right| [B_{DC} - \mu_0 M_S]$$
$$= 28 [0.5 - 0.176]$$
$$\boxed{f_\perp = 9.07 \text{ GHz}}$$

Problem 5.3 The uniform precession mode of ferrite spheres is widely used to make tunable microwave filters. What field B_{DC} is required to obtain a resonance at 10 GHz in YIG?

Solution 5.3

Since the resonant frequency of a sphere depends only on the strength of the applied field, we have

$$f_{up} = \left| \frac{\gamma}{2\pi} \right| B_{DC}$$
$$\boxed{\therefore \quad B_{DC} = \frac{10}{28} = 0.357 \text{ T}}$$

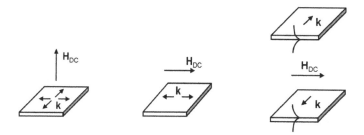

Fig. 5.2 The external field H_{DC} can be perpendicular or parallel to the film plane. The orientations of the applied field H_{DC} and the wave vector **k** determine the type of spin wave excitation, i.e., forward volume, backward volume, or surface wave (Reproduced from [4], with permission)

5.3 Dipolar Spin Waves

The form of the gyrotropic permeability tensor (5.14) results from precessing spins. Consequently, we can view a magnetostatic mode as a type of spin wave. When the wavelength is much larger than a lattice constant a but much shorter than that of an electromagnetic wave, neighboring spins are very nearly parallel, and the dynamic energy stored in the exchange interaction is negligible. Under these conditions, the long-range dipolar coupling between spins dominates the dynamics, and the waves are referred to as *dipolar spin waves.*

Electromagnetic waves in unbounded media are described by the solutions to Maxwell's equations. For the case of waveguides, we impose the boundary conditions

- tangential **H** must be continuous, and
- normal **B** must be continuous.

When we consider spin wave propagation in magnetic thin films, we look for solutions to Walker's equation (5.13) and also account for the direction of the applied external magnetic field, relative to the direction of spin wave propagation. This leads us to the three possible configurations, as shown in Fig. 5.2:

- *Forward volume waves*, in a normally magnetized film, with $\mathbf{k} \perp \mathbf{H}_{DC}$
- *Backward volume waves*, in a tangentially magnetized film, with $\mathbf{k} \parallel \mathbf{H}_{DC}$
- *Surface waves*, in a tangentially magnetized film, with $\mathbf{k} \perp \mathbf{H}_{DC}$

We will revisit the concept of a *forward* and a *backward wave* in Sect. 10.2.

5.3.1 Forward Volume Waves

We begin by considering the geometry shown in Fig. 5.3, with spin waves bouncing back and forth between the two surfaces of the film. Let us assume a symmetric superposition of bouncing plane waves in Region II as a trial solution. The magnetostatic potential takes the form

Fig. 5.3 The guided wave in a thin film can be visualized as being made up of plane waves bouncing back and forth between the top and bottom surfaces of the film (Reproduced from [4], with permission)

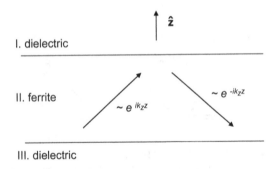

$$\psi_{\mathrm{II}} = \psi_0\, e^{i\mathbf{k}_t \cdot \mathbf{r}} \left[\frac{e^{ik_z z} + e^{-ik_z z}}{2}\right] = \psi_0\, \cos(k_z z)\, e^{i\mathbf{k}_t \cdot \mathbf{r}}, \qquad (5.35)$$

where ψ_0 is a chosen mode amplitude. In the dielectric, $\chi = 0$, and Walker's equation (5.13) reduces to *Laplace's equation*,

$$\nabla^2 \psi = 0. \qquad (5.36)$$

From (5.15) we have

$$k_{t,d}^2 + k_{z,d}^2 = 0 \implies k_{z,d} = \pm i k_{t,d}. \qquad (5.37)$$

Here the subscript d refers to the dielectric regions outside of the film. This suggests trial solutions of the form

$$\psi a(\mathbf{r}) = C e^{i\mathbf{k}_{t,d} \cdot \mathbf{r} \pm k_{t,d} z}. \qquad (5.38)$$

We choose the signs in the exponent so that the potential vanishes as $z \to \pm\infty$ to ensure that the mode is localized to the film:

$$\psi_{\mathrm{I}}(\mathbf{r}) = C e^{i\mathbf{k}_{t,d} \cdot \mathbf{r} - k_{t,d} z}, \qquad (5.39a)$$

$$\psi_{\mathrm{III}}(\mathbf{r}) = D e^{i\mathbf{k}_{t,d} \cdot \mathbf{r} + k_{t,d} z}. \qquad (5.39b)$$

We now apply the boundary conditions on tangential $\mathbf{h}$ and normal $\mathbf{b}$ at $z = \pm d/2$ to obtain two linearly independent equations

$$C e^{-k_t d/2} = \psi_0 \cos(k_z d/2), \qquad (5.40a)$$

$$k_t C e^{-k_t d/2} = \psi_0 k_z \sin(k_z d/2), \qquad (5.40b)$$

that we combine to get

$$\psi_0 k_t \cos(k_z d/2) = \psi_0 k_z \sin(k_z d/2) \qquad (5.41)$$

or

$$\tan(k_z d/2) = \frac{k_t}{k_z}. \qquad (5.42)$$

This equation is further simplified by recognizing that $(1 + \chi)k_t^2 + k_z^2 = 0$, from Walker's equation, and we obtain the dispersion relation for forward volume waves (FVWs) with symmetric thickness profiles (even modes) as

$$\tan\left[\frac{k_t d}{2}\sqrt{-(1+\chi)}\right] = \frac{1}{\sqrt{-(1+\chi)}}. \qquad (5.43)$$

After a similar analysis for forward volume waves with odd thickness profiles, we find that both even and odd mode dispersion relations can be expressed as

$$\tan\left[\frac{k_t d}{2}\sqrt{-(1+\chi)} - \frac{n\pi}{2}\right] = \frac{1}{\sqrt{-(1+\chi)}}, \quad n = 0, 1, 2 \dots. \qquad (5.44)$$

Here even values of n correspond to modes with symmetric thickness variations (even modes), and odd values of n correspond to modes with antisymmetric thickness variations (odd modes). For the lowest-order ($n = 0$) mode, an approximation to the solution of this *transcendental equation*, that can be solved explicitly for ω, is [5]

$$\omega^2 = \omega_0 \left[\omega_0 + \omega_M \left(1 - \frac{1 - e^{-k_t d}}{k_t d}\right)\right]. \qquad (5.45)$$

5.3.2   Backward Volume Waves

In a manner similar to that followed for forward volume waves, we solve for the magnetostatic potential ψ in the case of backward volume waves (BVWs) to obtain the dispersion relation

$$\tan\left[\frac{k_z d}{2\sqrt{-(1+\chi)}} - \frac{(n-1)\pi}{2}\right] = \sqrt{-(1+\chi)}, \quad n = 1, 2, 3 \dots. \qquad (5.46a)$$

As before, even values of n correspond to modes with symmetric thickness variations (even modes), and odd values of n correspond to modes with antisymmetric thickness variations (odd modes). However, in contrast to forward volume waves, the lowest-order backward volume wave mode has odd thickness symmetry. For the lowest-order mode ($n = 1$), we have the approximation [5]

$$\omega^2 = \omega_0 \left[\omega_0 + \omega_M \left(\frac{1 - e^{-k_z d}}{k_z d}\right)\right]. \qquad (5.46b)$$

5.3.3 Surface Waves

In the case of surface waves, $k_z = 0$ and the plane wave form of Walker's equation (5.15) requires that $k_y^2 = -k_x^2$ if $1 + \chi \neq 0$. As a result the potential in Region II takes the form

$$\psi_{\mathrm{II}} = \left[\psi_{0+} e^{k_y y} + \psi_{0-} e^{-k_y y} \right] e^{i \nu k_x x}, \tag{5.47}$$

where $\nu = \pm 1$ depending on the direction of propagation and k_x is taken to be positive. Note that we can also drop the wave number subscripts since $|k_y| = k_x \equiv k$. Following the procedure of applying the boundary conditions, we obtain

$$e^{-2kd} = \frac{(\chi + 2)^2 - \kappa^2}{\chi^2 - \kappa^2}, \tag{5.48a}$$

where dependence on ν has vanished, showing that the dispersion relation is unchanged if the direction of propagation is reversed. Although the dispersion relation is symmetric, a closer examination shows that the mode fields shift from one surface to the other depending on the direction of propagation, as suggested by the third panel in Fig. 5.2.

Substituting for χ and κ from (3.63), and solving for ω^2 gives

$$\omega^2 = \omega_0 (\omega_0 + \omega_M) + \frac{\omega_M^2}{4} \left[1 - e^{-2kd} \right]. \tag{5.48b}$$

We could also solve for k to get

$$k = -\frac{1}{2d} \ln \left[1 + \frac{4}{\omega_M^2} \left[\omega_0 (\omega_0 + \omega_M) - \omega^2 \right] \right]. \tag{5.49}$$

Problem 5.4 Using the *image theorem*, find the forward volume wave dispersion relation for the lowest mode of a ferrite film with one side in contact with a perfect conductor. Show that your result is identical to the even mode dispersion relation of an isolated film twice as thick.

Solution 5.4

In the forward volume wave geometry, small-signal $\overline{m}$ is tangential to the perfect conductor, as illustrated in Fig. 5.4. Since magnetic dipoles parallel to a perfect conductor are imaged without inversion, $\overline{m}$ and the potential ψ above the ground plane are the same as the top half of even modes in a film twice as thick.

In the absence of a ground plane, the dispersion relation is

$$\tan \left[\frac{k_t d}{2} \sqrt{-(1 + \chi)} - \frac{n \pi}{2} \right] = \frac{1}{\sqrt{-(1 + \chi)}}. \tag{5.44}$$

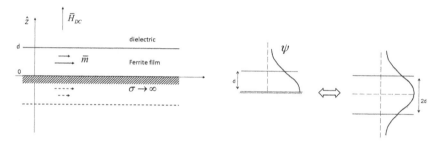

Fig. 5.4 The modes of a film in contact with a perfect ground plane are equivalent to the even modes of an isolated film twice as thick. Left: Magnetic dipoles are imaged above a perfect ground plane without inversion. Right: Magnetostatic potential function imaged above a perfect conductor

For the $n = 0$ mode of a film with thickness $2d$, this reduces to

$$\tan\left[k_t d\sqrt{-(1+\chi)}\right] = \frac{1}{\sqrt{-(1+\chi)}}.$$

To check, the dispersion relation for forward volume waves with a gap between the film and ground plane is given by (see Problem 5.5)

$$-\tan\left[k_t d\sqrt{-(1+\chi)}\right] = \frac{2\sqrt{-(1+\chi)}}{2 + \chi\left(1 + e^{-2k_t t}\right)},$$

where t is the distance to the ground plane. Setting $t = 0$ to put the ground plane against the film gives

$$-\tan\left[k_t d\sqrt{-(1+\chi)}\right] = \frac{2\sqrt{-(1+\chi)}}{2 + 2\chi}$$

$$= \frac{\sqrt{-(1+\chi)}}{1 + \chi}$$

$$\therefore \quad \tan\left[k_t d\sqrt{-(1+\chi)}\right] = \frac{1}{\sqrt{-(1+\chi)}}$$

This is the same equation obtained above using image theory.

Problem 5.5 The most common magnetostatic wave device geometry consists of a YIG film above a ground plane as shown in Fig. 5.5. Consider a layered geometry consisting of a perfect conductor for $z \leq -t$, a nonmagnetic dielectric for $-t < z \leq 0$, a ferrite film for $0 < z \leq d$, and a nonmagnetic dielectric for $z > d$. Show that the dispersion relation for forward volume waves can be written

$$-\tan\left[kd\sqrt{-(1+\chi)}\right] = \frac{2\sqrt{-(1+\chi)}}{2 + \chi(1 + e^{-2kt})}, \tag{5.50}$$

where k is the wave number in the plane of the film (the subscript t has been omitted for clarity).

Fig. 5.5 Common experimental layered structure with a ground plane. Forward volume waves propagate when the bias field is normal to the film plane

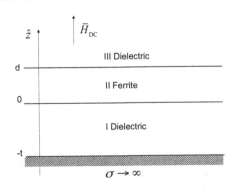

Solution 5.5

We begin with the trial solutions

$$\psi_I = \left(A_1 e^{kz} + A_2 e^{-kz}\right) e^{i\bar{k}\cdot\bar{r}}$$

$$\psi_{II} = (C_1 \cos k_z z + C_2 \sin k_z z)\, e^{i\bar{k}\cdot\bar{r}}$$

$$\psi_{III} = D e^{-kz} e^{i\bar{k}\cdot\bar{r}}$$

where $\bar{k}$ is a vector in the x-y plane. Requiring ψ to be continuous gives the equations:

$$\psi_{II}(0) = \psi_I(0) \Rightarrow C_1 = A_1 + A_2 \tag{5.51a}$$

$$\psi_{III}(d) = \psi_{II}(d) \Rightarrow D e^{-kd} = C_1 \cos k_z d + C_2 \sin k_z d \tag{5.51b}$$

Since there is no small-signal susceptibility along $\hat{z}$, continuity of normal $\bar{b}$ reduces to the requirement that $\dfrac{\partial \psi}{\partial z}$ be continuous at the interface:

$$\left.\frac{\partial \psi_I}{\partial z}\right|_{z=-t} = 0 \quad \Longrightarrow \quad A_1 e^{-kt} - A_2 e^{kt} = 0 \tag{5.52a}$$

$$\left.\frac{\partial \psi_I}{\partial z}\right|_{z=0} = \left.\frac{\partial \psi_{II}}{\partial z}\right|_{z=0} \quad \Longrightarrow \quad k(A_1 - A_2) = k_z C_2 \tag{5.52b}$$

$$\left.\frac{\partial \psi_{II}}{\partial z}\right|_{z=d} = \left.\frac{\partial \psi_{III}}{\partial z}\right|_{z=d} \quad \Longrightarrow \quad k_z(-C_1 \sin k_z d + C_2 \cos k_z d) = -kD e^{-kd} \tag{5.52c}$$

We now combine (5.51a), (5.52a), and (5.52b) to eliminate A_1, A_2.

$$C_1 = A_2 \left(1 + e^{2kt}\right)$$

$$k_z C_2 = k A_2 \left(e^{2kt} - 1\right)$$

$$= k \left(e^{2kt} - 1\right) \left[\frac{C_1}{1 + e^{2kt}}\right]$$

$$\therefore \quad k_z \left(1 + e^{2kt}\right) C_2 = k \left(e^{2kt} - 1\right) C_1 \tag{5.53}$$

From (5.51b) and (5.52c):

$$k_z \left[-C_1 \sin k_z d + C_2 \cos k_z d\right] = -k \left[C_1 \cos k_z d + C_2 \sin k_z d\right]$$

or,

$$C_1 \left[k \cos k_z d - k_z \sin k_z d\right] + C_2 \left[k_z \cos k_z d + k \sin k_z d\right] = 0 \tag{5.54}$$

Combining (5.53) and (5.54):

$$\frac{k_z \left(1 + e^{2kt}\right)}{k \left(e^{2kt} - 1\right)} \left[k \cos k_z d - k_z \sin k_z d\right] + \left[k_z \cos k_z d + k \sin k_z d\right] = 0$$

$$\left(1 + e^{2kt}\right) \left[k_z \cos k_z d - \frac{k_z^2}{k} \sin k_z d\right] + \left(e^{2kt} - 1\right) \left[k_z \cos k_z d + k \sin k_z d\right] = 0$$

$$k_z \cos k_z d \left[1 + e^{2kt} + e^{2kt} - 1\right] + k \sin k_z d \left[\left(e^{2kt} - 1\right) - \frac{k_z^2}{k^2}\left(1 + e^{2kt}\right)\right] = 0$$

$$2e^{2kt} + \frac{k}{k_z} \tan k_z d \left[e^{2kt}\left(1 - \frac{k_z^2}{k^2}\right) - \left(1 + \frac{k_z^2}{k^2}\right)\right] = 0$$

From Walker's Equation (5.13)

$$\frac{k_z}{k} = \sqrt{-(1 + \chi)}. \tag{5.55}$$

Hence,

$$2e^{2kt} + \frac{1}{\sqrt{-(1+\chi)}} \tan k_z d \left[e^{2kt}(1 + 1 + \chi) - (1 - 1 - \chi)\right] = 0$$

$$2e^{2kt} + \frac{1}{\sqrt{-(1+\chi)}} \tan k_z d \left[e^{2kt}(2 + \chi) + \chi\right] = 0$$

$$2e^{2kt} + \frac{1}{\sqrt{-(1+\chi)}} \tan k_z d \left[\chi \left(1 + e^{2kt}\right) + 2e^{2kt}\right] = 0$$

$$2 + \frac{1}{\sqrt{-(1+\chi)}} \tan k_z d \left[\chi \left(e^{-2kt} + 1\right) + 2\right] = 0$$

Fig. 5.6 Layered structure backed by a ground plane often used in experiments when a magnetic film is placed in contact with a microstrip circuit. Backward volume waves propagate when the wave vector and the bias field are parallel and in the plane of the film

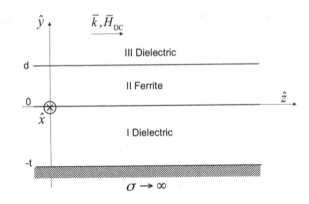

Rearranging terms:

$$\tan k_z d = \frac{-2\sqrt{-(1+\chi)}}{2 + \chi\left(1 + e^{-2kt}\right)} \quad \Rightarrow \quad \boxed{\tan\sqrt{-(1+\chi)}kd = \frac{-2\sqrt{-(1+\chi)}}{2 + \chi\left(1 + e^{-2kt}\right)}.}$$

Problem 5.6 Consider magnetostatic backward volume waves in a ferrite film with an adjacent ground plane. The geometry consists of a perfect conductor for $y \leq -t$, a nonmagnetic dielectric for $-t < y \leq 0$, a ferrite film for $0 < y \leq d$ and a nonmagnetic dielectric for $y > d$, as shown in Fig. 5.6. Show that the dispersion relation for backward volume waves can be written

$$\tan\left[\frac{kd}{\sqrt{-(1+\chi)}}\right] = \frac{2\sqrt{-(1+\chi)}}{2 + \chi(1 + e^{-2kt})}, \tag{5.56}$$

where k is the wave number along the direction of propagation (the subscript z has been omitted for clarity).

Solution 5.6

For backward volume spin waves with a ground plane, we assume $k_y = k_z = k$ in regions I and III (as per Laplace's equation, (5.36)) and start with the trial solutions

$$\psi_\mathrm{I} = \left(A_1 e^{ky} + A_2 e^{-ky}\right) e^{ikz}$$

$$\psi_\mathrm{II} = \left(C_1 \cos k_y y + C_2 \sin k_y y\right) e^{ikz}$$

$$\psi_\mathrm{III} = D e^{-ky} e^{ikz}$$

Requiring ψ to be continuous gives the equations:

$$\psi_\mathrm{II}(0) = \psi_\mathrm{I}(0) \quad \Longrightarrow \quad C_1 = A_1 + A_2 \tag{5.57a}$$

$$\psi_\mathrm{III}(d) = \psi_\mathrm{II}(d) \quad \Longrightarrow \quad D e^{-kd} = C_1 \cos k_y d + C_2 \sin k_y d \tag{5.57b}$$

In region II, $b_y = -\mu_0 (1 + \chi) \dfrac{\partial \psi_{\text{II}}}{\partial y}$ while in regions I and III, $b_y = -\mu_0 \dfrac{\partial \psi_{\text{I,III}}}{\partial y}$.

Therefore the boundary conditions from normal $\overline{b}$ continuous give

$$\left. \frac{\partial \psi_{\text{I}}}{\partial y} \right|_{y=0} = (1 + \chi) \left. \frac{\partial \psi_{\text{II}}}{\partial y} \right|_{y=0}$$

$$\Rightarrow kA_1 - kA_2 = (1 + \chi) k_y \left[-C_1 \sin k_y y + C_2 \cos k_y y \right]|_{y=0}$$

$$k(A_1 - A_2) = (1 + \chi) k_y C_2$$

But from Walker's equation (5.13), we have $\dfrac{k_y}{k} = \dfrac{1}{\sqrt{-(1 + \chi)}}$

$$\Rightarrow A_1 - A_2 = \frac{(1 + \chi)}{\sqrt{-(1 + \chi)}} C_2$$

$$-A_1 + A_2 = \sqrt{-(1 + \chi)} C_2 \tag{5.58}$$

Similarly, we have boundary conditions at $y = d$,

$$(1 + \chi) \left. \frac{\partial \psi_{\text{II}}}{\partial y} \right|_{y=d} = \left. \frac{\partial \psi_{\text{III}}}{\partial y} \right|_{y=d}$$

that yield,

$$(1 + \chi) k_y \left[-C_1 \sin k_y d + C_2 \cos k_y d \right] = -kDe^{-kd}$$

$$-(1 + \chi) \frac{k_y}{k} \left[-C_1 \sin k_y d + C_2 \cos k_y d \right] = De^{-kd}$$

Therefore,

$$\sqrt{-(1 + \chi)} \left[-C_1 \sin k_y d + C_2 \cos k_y d \right] = De^{-kd}. \tag{5.59}$$

Finally, at $y = t$, $b_y = 0$ gives

$$\left. \frac{\partial \psi_{\text{I}}}{\partial y} \right|_{y=-t} = 0 = k \left(A_1 e^{-kt} - A_2 e^{kt} \right)$$

or

$$A_1 = A_2 e^{2kt} \tag{5.60}$$

Combining (5.57a) and (5.60):

$$C_1 = \left(1 + e^{2kt} \right) A_2 \tag{5.61}$$

Combining (5.58) and (5.60):

$$C_2\sqrt{-(1+\chi)} = \left(1 - e^{2kt}\right) A_2 \tag{5.62}$$

From (5.57b) and (5.59):

$$C_1 \cos k_y d + C_2 \sin k_y d = \sqrt{-(1+\chi)}\left(-C_1 \sin k_y d + C_2 \cos k_y d\right)$$
$$\cos k_y d \left[C_1 + C_2 \tan k_y d\right] = \sqrt{-(1+\chi)}\left[C_2 - C_1 \tan k_y d\right] \cos k_y d.$$

If $\cos k_y d \neq 0$, then

$$C_1 + C_2 \tan k_y d = \sqrt{-(1+\chi)}\left[C_2 - C_1 \tan k_y d\right]$$
$$\tan k_y d \left[C_2 + C_1\sqrt{-(1+\chi)}\right] = C_2\sqrt{-(1+\chi)} - C_1 \tag{5.63}$$

From (5.61) and (5.62):

$$C_2\sqrt{-(1+\chi)} = \frac{\left(1 - e^{2kt}\right) C_1}{\left(1 + e^{2kt}\right)} \tag{5.64}$$

Combining (5.63) and (5.64):

$$\tan k_y d \left[\frac{1}{\sqrt{-(1+\chi)}}\frac{\left(1 - e^{2kt}\right)}{1 + e^{2kt}} + \sqrt{-(1+\chi)}\right] = \frac{1 - e^{2kt}}{1 + e^{2kt}} - 1$$

$$\tan k_y d \left[\frac{1 - e^{2kt} - \left(1 + e^{2kt}\right)(1+\chi)}{\sqrt{-(1+\chi)}\left(1 + e^{2kt}\right)}\right] = \frac{1 - e^{2kt} - 1 - e^{2kt}}{1 + e^{2kt}}$$

$$\tan k_y d \left[\frac{1 - e^{2kt} - \left(1 + e^{2kt}\right) - \chi\left(1 + e^{2kt}\right)}{\sqrt{-(1+\chi)}}\right] = -2e^{2kt}$$

$$\tan k_y d \left[\frac{-2e^{2kt} - \chi\left(1 + e^{2kt}\right)}{\sqrt{-(1+\chi)}}\right] = -2e^{2kt}$$

$$\tan k_y d \left[\frac{2 + \chi\left(1 + e^{-2kt}\right)}{\sqrt{-(1+\chi)}}\right] = 2$$

Solving for $\tan k_y d$ and using Walker's equation (5.13) gives (5.56)

$$\tan\left[\frac{kd}{\sqrt{-(1+\chi)}}\right] = \frac{2\sqrt{-(1+\chi)}}{2 + \chi\left(1 + e^{-2kt}\right)}.$$

Fig. 5.7 Layered structure similar to Fig. 5.6 except that propagation is along $\pm x$ instead of z. This geometry supports surface waves

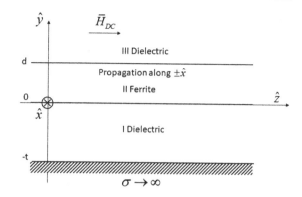

Problem 5.7 Consider magnetostatic surface waves in a ferrite film above a ground plane. Use the geometry of Fig. 5.7 and show that the dispersion relation can be written

$$e^{2kd} = \frac{\chi + \nu\kappa}{2 + \chi - \nu\kappa}\left[\frac{\chi - \nu\kappa + e^{-2kt}(2 + \chi - \nu\kappa)}{2 + \chi + \nu\kappa + e^{-2kt}(\chi + \nu\kappa)}\right]. \tag{5.65}$$

Solution 5.7

For surface wave propagation with a ground plane, we assume the trial solutions

$$\psi_{\text{I}} = \left(A_1 e^{ky} + A_2 e^{-ky}\right)e^{i\nu kx}, \ \ \nu = \pm 1$$
$$\psi_{\text{II}} = \left(C_1 e^{ky} + C_2 e^{-ky}\right)e^{i\nu kx}$$
$$\psi_{\text{III}} = D e^{-ky}e^{i\nu kx}$$

Requiring ψ to be continuous gives

$$\psi_{\text{I}}(0) = \psi_{\text{II}}(0) \Rightarrow A_1 + A_2 = C_1 + C_2 \tag{5.66a}$$
$$\psi_{\text{II}}(d) = \psi_{\text{III}}(d) \Rightarrow C_1 e^{kd} + C_2 e^{-kd} = D e^{-kd} \tag{5.66b}$$

For region II,

$$b_y = -i\mu_0\kappa\frac{\partial\psi_{\text{II}}}{\partial x} - \mu_0(1 + \chi)\frac{\partial\psi_{\text{II}}}{\partial y},$$

while in regions I and III

$$b_y = -\mu_0\frac{\partial\psi_{\text{I,III}}}{\partial y}.$$

Therefore the boundary conditions from normal **b** continuous give

$$
\left.\frac{\partial \psi_{\mathrm{I}}}{\partial y}\right|_{y=0} = \left[i\kappa \frac{\partial \psi_{\mathrm{II}}}{\partial x} + (1+\chi)\frac{\partial \psi_{\mathrm{II}}}{\partial y}\right]_{y=0}
$$
$$
(A_1 - A_2) = -\nu\kappa\,(C_1 + C_2) + (1+\chi)\,(C_1 - C_2)
$$
$$
(A_1 - A_2) = C_1\,(-\nu\kappa + (1+\chi)) + C_2\,(-\nu\kappa - (1+\chi)) \qquad (5.67)
$$

and

$$
\left.\frac{\partial \psi_{\mathrm{III}}}{\partial y}\right|_{y=d} = \left[i\kappa \frac{\partial \psi_{\mathrm{II}}}{\partial x} + (1+\chi)\frac{\partial \psi_{\mathrm{II}}}{\partial y}\right]_{y=d}
$$
$$
-De^{-kd} = -\nu\kappa\left(C_1 e^{kd} + C_2 e^{-kd}\right) + (1+\chi)\left(C_1 e^{kd} - C_2 e^{-kd}\right)
$$
$$
-De^{-kd} = C_1 e^{kd}\,[-\nu k + (1+\chi)] + C_2 e^{-kd}\,[-\nu k - (1+\chi)] \qquad (5.68)
$$

Finally, at $y = -t$, $b_y = 0$ gives

$$
\left.\frac{\partial \psi_{\mathrm{I}}}{\partial y}\right|_{y=-t} = 0 = k\left(A_1 e^{-kt} - A_2 e^{kt}\right)
$$

or,

$$
A_1 e^{-kt} = A_2 e^{kt}. \qquad (5.69)
$$

Combining (5.66a) and (5.69):

$$
A_1\left(1 + e^{-2kt}\right) = C_1 + C_2 \qquad (5.70)
$$

Combining (5.67) and (5.69):

$$
A_1\left(1 - e^{-2kt}\right) = C_1\,[-\nu k + 1 + \chi] + C_2\,[-\nu\kappa - 1 - \chi] \qquad (5.71)
$$

Combining (5.70) and (5.71):

$$
(C_1 + C_2)\frac{\left(1 - e^{-2kt}\right)}{\left(1 + e^{-2kt}\right)} = C_1\,[-\nu\kappa + 1 + \chi] + C_2\,[-\nu\kappa - 1 - \chi]
$$
$$
(C_1 + C_2)\left(1 - e^{-2kt}\right) = C_1\left(1 + e^{-2kt}\right)[-\nu\kappa + 1 + \chi]
$$
$$
+ C_2\left(1 + e^{-2kt}\right)[-\nu\kappa - 1 - \chi]
$$

We rewrite this as

$$
C_1\left[-\nu\kappa + 1 + \chi + e^{-2kt}\,(-\nu\kappa + 1 + \chi) - 1 + e^{-2kt}\right]
$$
$$
+ C_2\left[-\nu\kappa - 1 - \chi + e^{-2kt}\,(-\nu\kappa - 1 - \chi) - 1 + e^{-2kt}\right] = 0
$$

or,

$$C_1\left[-\nu\kappa + \chi + e^{-2kt}\left(-\nu\kappa + 2 + \chi\right)\right]$$
$$+ C_2\left[-\nu\kappa - 2 - \chi + e^{-2kt}\left(-\nu\kappa - \chi\right)\right] = 0. \tag{5.72}$$

Combining (5.66b) and (5.68) gives:

$$-C_1 e^{-kd} - C_2 e^{-kd} = C_1 e^{kd}\left[-\nu\kappa + 1 + \chi\right] + C_2 e^{-kd}\left[-\nu\kappa - 1 - \chi\right]$$

or,

$$C_1 e^{kd}\left[-\nu\kappa + 2 + \chi\right] + C_2 e^{-kd}\left[-\nu\kappa - \chi\right] = 0. \tag{5.73}$$

Equations (5.72) and (5.73) can be combined to obtain the matrix equation

$$\begin{bmatrix} -\nu\kappa + \chi + e^{-2kt}\left(-\nu\kappa + 2 + \chi\right) - \left(\nu\kappa + 2 + \chi\right) - e^{-2kt}\left[\nu\kappa + \chi\right] \\ e^{kd}\left[-\nu\kappa + 2 + \chi\right] \qquad\qquad -e^{-kd}\left[\nu\kappa + \chi\right] \end{bmatrix}\begin{bmatrix} C_1 \\ C_2 \end{bmatrix} = 0$$

Setting the determinant to zero yields

$$- e^{-kd}\left(\nu\kappa + \chi\right)\left[-\nu\kappa + \chi + e^{-2kt}\left(-\nu\kappa + 2 + \chi\right)\right]$$
$$+ e^{kd}\left(-\nu\kappa + 2 + \chi\right)\left[\nu\kappa + 2 + \chi + e^{-2kt}\left(\nu\kappa + \chi\right)\right] = 0$$

or,

$$e^{2kd}\left(-\nu\kappa + 2 + \chi\right)\left[\nu\kappa + 2 + \chi + e^{-2kt}\left(\nu\kappa + \chi\right)\right]$$
$$= \left(\nu\kappa + \chi\right)\left[-\nu\kappa + \chi + e^{-2kt}\left(-\nu\kappa + 2 + \chi\right)\right].$$

Thus, we finally get (5.65),

$$\boxed{e^{2kd} = \frac{\left(\nu\kappa + \chi\right)}{\left(2 + \chi - \nu\kappa\right)}\frac{\left[\chi - \nu\kappa + e^{-2kt}\left(2 + \chi - \nu\kappa\right)\right]}{\left[2 + \chi + \nu\kappa + e^{-2kt}\left(\chi + \nu\kappa\right)\right]}.}$$

5.4 Exchange-Dominated Spin Waves

When exchange cannot be neglected, $\mathbf{h}$ is obtained from $\mathbf{m}$ using the matrix differential operator $\overline{\mathbf{A}}_{\text{op}}$

$$\mathbf{h} = \overline{\mathbf{A}}_{\text{op}} \cdot \mathbf{m} \tag{5.74}$$

where

$$\overline{\mathbf{A}}_{\text{op}} = \frac{1}{\omega_{\text{M}}}\begin{bmatrix} \omega_0 - \omega_{\text{M}}\lambda_{\text{ex}}\nabla^2 & i\omega \\ -i\omega & \omega_0 - \omega_{\text{M}}\lambda_{\text{ex}}\nabla^2 \end{bmatrix} \tag{5.75}$$

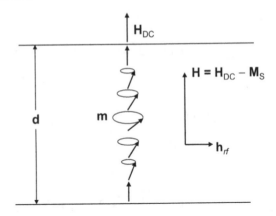

Fig. 5.8 Standing spin wave resonance in a magnetic thin film (Reproduced from [4], with permission)

and we have neglected magnetocrystalline anisotropy.[5] For a uniform plane wave, the operator ∇^2 can be replaced by the factor $-k^2$. Looking at the form of $\overline{\mathbf{A}}_{op}$, we observe that ω_0 becomes $\omega_0 + \omega_M \lambda_{ex} k^2$, and the dispersion relation (5.19) becomes

$$\omega = \left[(\omega_0 + \omega_M \lambda_{ex} k^2)(\omega_0 + \omega_M (\lambda_{ex} k^2 + \sin^2 \theta))\right]^{1/2}. \tag{5.76}$$

The value of the exchange constant determined from spin wave measurements at microwave frequencies in YIG is $\lambda_{ex} = 3 \times 10^{-16}\,\text{m}^2$ [6].

Seavey and Tannenwald [7] observed resonances in thin films that could be explained if the spins near the surfaces behaved as if they were "pinned" and not allowed to precess. In this case *spin wave resonances* would occur whenever the film thickness equaled an integral number of half wavelengths (Fig. 5.8). The resonant frequencies would then be given by (5.76) with $k = n\pi/d$ where d is the film thickness and n is an integer.

For a film magnetized normal to its plane, the static field in the film is reduced by the presence of the demagnetizing field, $-M_S$, generated by effective magnetic charges on the film surfaces (see Fig. 5.1). Setting $\theta = 0$ in (5.76) we get

$$\omega_n = \omega_0 + \omega_M \lambda_{ex} \left(\frac{\pi n}{d}\right)^2, \tag{5.77}$$

where

$$\omega_0 = -\gamma \mu_0 (H_{DC} - M_S). \tag{5.78}$$

Problem 5.8 Calculate the frequencies of the spin wave resonance modes for $n = 1, 3, 5, 7, 9$ in a $0.5\,\mu\text{m}$ YIG film if the field $H_{DC} = 240\,\text{kA/m}$ is applied normal to the film plane. For YIG, $M_S = 140\,\text{kA/m}$, $|\gamma|/2\pi = 28\,\text{GHz/T}$, and $\lambda_{ex} = 3 \times 10^{-16}\,\text{m}^2$.

---

[5]A derivation of the linearized torque equation, with both exchange and anisotropy, is given in Stancil and Prabhakar [4, Sect. 3.7.3].

Table 5.1 Spin wave resonance frequencies of a thin film magnetized normal to the plane of the film

n	1	3	5	7	9
f_n(GHz)	3.577	4.044	4.978	6.378	8.245

Solution 5.8

Assuming that we have standing waves across the thickness of the film, we get

$$\omega_n = \omega_0 + \omega_M \lambda_{ex} \left(\frac{n\pi}{d}\right)^2$$

or

$$f_n = f_0 + f_M \lambda_{ex} \left(\frac{n\pi}{d}\right)^2.$$

Since the DC field is normal to the film plane, the demagnetizing factor is unity, and the demagnetizing field is equal to the magnetization and opposite to the applied DC filed. Thus using the given values of H_{DC} and M_S,

$$f_0 = \left|\frac{\gamma}{2\pi}\right| \mu_0 (H_{DC} - M_S) = 28 (4\pi) 10^{-7} 100 \left(10^3\right)$$

$$= 3.519\,\text{GHz}$$

$$f_M = \left|\frac{\gamma}{2\pi}\right| \mu_0 M_S = 28 (4\pi) 10^{-7} 140 \left(10^3\right)$$

$$= 4.926\,\text{GHz}$$

The resonant frequencies are then calculated to be (Table 5.1)

$$f_n = 3.519 + 4.926 (3) 10^{-16} \left(\frac{n\pi}{0.5 \left(10^{-6}\right)}\right)^2$$

$$= 3.519 + 0.0583\, n^2.$$

References

1. B. Lax, K.J. Button, *Microwave Ferrites and Ferrimagnetics* (McGraw-Hill, New York, 1962)
2. R.F. Sooho, *Microwave Magnetics* (Harper and Row, New York, 1985)
3. M.S. Sodha, N.C. Srivastava, *Microwave Propagation in Ferrimagnetics* (Harper and Row, New York, 1981)
4. D.D. Stancil, A. Prabhakar, *Spin Waves: Theory and Applications* (Springer, New York, 2009)
5. B.A. Kalinikos, Excitation of propagating spin waves in ferromagnetic films. IEE Proc. **127**(H), 4–10 (1980)
6. E.H. Turner, Interaction of phonons and spin waves in yttrium iron garnet. Phys. Rev. Lett. **5**, 100 (1960)
7. M.H. Seavey, P.E. Tannenwald, Direct observation of spin-wave resonance. Phys. Rev. Lett. **1**, 168–169 (1958)

Propagation Characteristics and Excitation of Dipolar Spin Waves

6

In this chapter, we expand on the properties of dipolar spin waves in thin films and describe how to excite them. The relaxation time and resulting attenuation of spin wave modes is explored in Problems 6.1–6.3. Mode orthogonality and normalization are considered in Problems 6.4–6.6.

We next consider the excitation of spin waves by a current filament. The power converted into spin waves manifests as an effective radiation resistance at the terminals of the current filament. The expressions for radiation resistance of volume waves are considered in Problems 6.7–6.8, while the radiation resistance for surface waves is considered in Problem 6.9.

6.1   Relaxation Time

We begin with a phenomenological model to describe propagation loss.[1] The equation of motion for the magnetization, including damping, is

$$\frac{d\mathbf{M}}{dt} = \gamma\mu_0\mathbf{M} \times \mathbf{H}_{\text{eff}} + \frac{\alpha}{M_S}\mathbf{M} \times \frac{d\mathbf{M}}{dt}, \tag{6.1}$$

where $\mathbf{H}_{\text{eff}}$ is the total effective internal magnetic field intensity. This is referred to as the Landau–Lifshitz–Gilbert (LLG) equation. In the following discussions, we will assume that $\mathbf{H}_{\text{eff}}$ is given completely by the applied field plus the demagnetizing field (exchange and magnetocrystalline anisotropy will be neglected for simplicity). In Chap. 3, we argued that the damping term in (6.1) can be included in the magnetic susceptibility tensor by allowing $\omega_0 \to \omega_0 - i\alpha\omega$, where $\omega_0 = -\gamma\mu_0 H_{\text{eff}}$. Inherent

---

[1]The treatment in this section follows Stancil [1].

© The Author(s), under exclusive license to Springer Nature Switzerland AG 2021
D. D. Stancil and A. Prabhakar, *Spin Waves*,
https://doi.org/10.1007/978-3-030-68582-9_6

in this argument was the assumption of a time dependence of the form $\exp(-i\omega t)$. Let us now define the *relaxation time* of a mode as the time required for the amplitude of the small-signal magnetization to decay by the factor $1/e$ after removal of an excitation. In an infinite medium, the relaxation time T_0 of the uniform precession mode is related to the damping parameter α as (see (3.76))

$$\frac{1}{T_0} = \omega\alpha, \tag{6.2}$$

where ω is the resonant frequency of the uniform precession mode. We can also relate T_0 to the full FMR line width ΔH of the uniform precession mode as (see (3.75))

$$\frac{1}{T_0} = -\frac{\gamma\mu_0\Delta H}{2}. \tag{6.3}$$

This allows us to relate α and ΔH

$$\alpha = -\frac{\gamma\mu_0\Delta H}{2\omega}. \tag{6.4}$$

Let us now assume a relaxation time T_k for a particular mode and geometry. Assuming the small-signal magnetization decays as $\exp(-t/T_k)$, we can define the loss per unit time

$$\begin{aligned} L &= -\frac{20\log_{10} e^{-t/T_k}}{t \times 10^6} \\ &= \frac{8.686}{T_k} \times 10^{-6}(\text{dB}/\mu\text{s}), \end{aligned} \tag{6.5}$$

where t and T_k are given in seconds.

If the time dependence is of the form $e^{-i\omega t}$, and we allow the frequency ω to be complex, then the amplitude will decay according to the factor $\exp(\text{Im}\{\omega\}t)$, where $\text{Im}\{\omega\} < 0$. A comparison with the loss parameter $\exp(-t/T_k)$ suggests the definition

$$\frac{1}{T_k} = -\text{Im}\{\omega\}. \tag{6.6}$$

Knowing that ω generally depends on ω_0, we write the Taylor series

$$\omega(\omega_0 + \Delta\omega_0) = \omega(\omega_0) + \frac{\partial\omega}{\partial\omega_0}\Delta\omega_0 + \ldots \tag{6.7}$$

In the presence of loss, we set $\Delta\omega_0 = -i/T_0$ to obtain

$$\omega(\omega_0 - i/T_0) \approx \omega(\omega_0) - \frac{\partial\omega}{\partial\omega_0}\frac{i}{T_0}, \tag{6.8}$$

to lowest order in $1/T_0$. Note that the first term in the Taylor series, $w(w_0)$, is the frequency in the absence of loss and is therefore real. Hence, we can approximate the imaginary part of w

$$\text{Im}\{w\} \approx \frac{-1}{T_0} \frac{\partial w}{\partial w_0}. \tag{6.9}$$

Combining this result with (6.6) gives

$$\frac{1}{T_k} = \frac{1}{T_0} \frac{\partial w}{\partial w_0}. \tag{6.10}$$

In general, however, spin wave dispersion relations are transcendental and of the form

$$F(w, k, w_0) = 0. \tag{6.11}$$

When it is not possible to solve explicitly for $w(k, w_0)$, the derivative in (6.10) is best obtained by implicit differentiation. The result is

$$\frac{1}{T_k} = -\frac{1}{T_0} \frac{\partial F/\partial w_0}{\partial F/\partial w}, \tag{6.12}$$

where the right-hand side is understood to be evaluated at $w_0 = -\gamma \mu_0 H_{\text{eff}}$.

We can now find expressions for T_k using the dispersion relations obtained in Chap. 5.

6.1.1 Surface Waves

The dispersion relation for surface dipolar spin waves in an isolated slab is given by (5.48b). Even though it is possible to solve for w in this case, it is computationally convenient to use (6.12) and rewrite (5.48b) as

$$F = \Gamma - \frac{w_{\text{M}}^2}{4} \left(1 - e^{-2kd}\right) = 0, \tag{6.13}$$

where

$$\Gamma = w^2 - w_0(w_0 + w_{\text{M}}). \tag{6.14}$$

Since Γ contains all the explicit dependence on w and w_0, we have

$$\frac{1}{T_k} = -\frac{1}{T_0} \frac{\partial \Gamma/\partial w_0}{\partial \Gamma/\partial w} = \frac{1}{T_0} \left[\frac{2w_0 + w_{\text{M}}}{2w}\right]. \tag{6.15}$$

We can also write this in terms of α or ΔH

$$\frac{1}{T_k} = \alpha \left(w_0 + \frac{w_{\text{M}}}{2}\right), \tag{6.16a}$$

$$\frac{1}{T_k} = \frac{|\gamma|\mu_0 \Delta H}{4w} (2w_0 + w_{\text{M}}). \tag{6.16b}$$

6.1.2 Volume Waves

The dispersion relations for forward and backward volume waves are given by (5.44) and (5.46a), respectively. A significant point to notice about both of these equations is that the dependence on ω and ω_0 enters only through χ. Because of this, we can write

$$\frac{\partial F}{\partial \omega} = \frac{\partial F}{\partial \chi}\frac{\partial \chi}{\partial \omega}, \tag{6.17}$$

$$\frac{\partial F}{\partial \omega_0} = \frac{\partial F}{\partial \chi}\frac{\partial \chi}{\partial \omega_0}. \tag{6.18}$$

Fortunately, the complicated quantity $\partial F/\partial \chi$ divides out of (6.12), leaving us with the much simpler result

$$\frac{1}{T_k} = -\frac{1}{T_0}\frac{\partial \chi/\partial \omega_0}{\partial \chi/\partial \omega}. \tag{6.19}$$

Evaluating the indicated derivatives in (6.19) using the definition of χ from (3.63a) gives

$$\frac{1}{T_k} = \frac{1}{T_0}\left[\frac{\omega_0^2 + \omega^2}{2\omega_0\omega}\right]. \tag{6.20}$$

Substituting for T_0, we obtain an expression that is valid for both forward and backward volume waves

$$\frac{1}{T_k} = \alpha\frac{\omega_0^2 + \omega^2}{2\omega_0} = |\gamma|\mu_0\Delta H\frac{\omega_0^2 + \omega^2}{4\omega_0\omega}. \tag{6.21}$$

Problem 6.1 A surface spin wave propagates in an isolated YIG film with $f = 6.3\,\text{GHz}$ and $B_{DC} = 0.15\,\text{T}$. The ferromagnetic resonance line width (full width, half maximum) is $\Delta H = 80\,\text{A/m}$ at 9.3 GHz. The gyromagnetic ratio for the film is 2.8 MHz/G.

(a) Find the Gilbert damping parameter, α.
(b) Calculate the loss per unit time, L, in dB/μs.

Solution 6.1

Given that $f = 6.3$ GHz, $B_{DC} = 0.15$ T and $\Delta H = 80$ A/m at 9.3 GHz, with $|\gamma/(2\pi)| = 28$ GHz/T

(a) We use (6.21) to calculate the damping constant

$$\alpha = \frac{|\gamma|\mu_0\Delta H}{2\omega} = \frac{|\gamma/(2\pi)|\mu_0\Delta H}{2f} = \frac{28(10^9)\,4\pi(10^{-7})80}{2(9.3)10^9}$$

$$\therefore \quad \boxed{\alpha = 1.51 \times 10^{-4}}$$

(b) The loss is given by (6.16a) combined with (6.5)

$$L = \alpha(\omega_0 + \omega_M/2)8.686(10^{-6}) \text{ dB/}\mu\text{s}.$$

Here

$$\omega_0 = |\gamma|B = 2\pi \times 28(10^9) \times 0.15 = 2.64(10^{10}),$$

and

$$\omega_M = |\gamma|\mu_0 M_S = 2\pi(28)10^9 4\pi(10^{-7})140(10^3) = 3.1(10^{10}).$$

Here we have used[2] $M_S = 140$ kA/m. Putting it all together gives

$$L = 1.51(10^{-4})\,(2.64 + 3.10/2)\,8.686(10^{-6})$$

$$\therefore \quad \boxed{L = 55 \quad \text{dB/}\mu\text{s}}$$

Problem 6.2 The ferromagnetic resonance line width (full width, half maximum) of a particular YIG film is $\Delta H = 60$ A/m at 9.3 GHz. Consider a forward volume spin wave propagating with $f = 3.9$ GHz and $H_{DC} = 239$ kA/m. The gyromagnetic ratio for the film is 2.8 MHz/G.

(a) Find the Gilbert damping parameter, α.
(b) Calculate the loss per unit time, L, in dB/μs.

Solution 6.2

(a) We use (6.21) to calculate the damping constant with $\Delta H = 60$ A/m at 9.3 GHz

$$\alpha = \frac{|\gamma|\,\mu_0 \Delta H}{2\omega} = \frac{|\frac{\gamma}{2\pi}|\mu_0 \Delta H}{2f}, \qquad \left|\frac{\gamma}{2\pi}\right| = 28 \quad \text{GHz/T}$$

$$= \frac{28\,(10^9)\,4\pi\,(10^{-7})\,60}{2\,(9.3)\,(10^9)}$$

$$\therefore \quad \boxed{\alpha = 1.14\,(10^{-4})}$$

---

[2]See Appendix A in Stancil and Prabhakar [2], for a full list of material properties.

(b) For volume waves, the relaxation time is given by (6.21)

$$\frac{1}{T_k} = \alpha \frac{(\omega_0^2 + \omega^2)}{2\omega_0}.$$

For forward volume waves

$$\begin{aligned}
\omega_0 &= |\gamma| \mu_0 (H_{DC} - M_S) \\
&= 2\pi \left(28 \times 10^9\right) \left(4\pi \times 10^{-7}\right) (239 - 140) \, 10^3 \\
&= 2.19 \left(10^{10}\right)
\end{aligned}$$

and

$$\omega = 2\pi \times 3.9 \left(10^9\right) = 2.45 \left(10^{10}\right)$$

$$\Rightarrow \quad \frac{1}{T_k} = 1.14 \frac{(10^{-4})}{2} \left[2.19 \left(10^{10}\right) + \frac{[2.45 \left(10^{10}\right)]^2}{2.19 \left(10^{10}\right)} \right] = 2.81 \left(10^6\right)$$

The loss/time is now given by

$$\begin{aligned}
L &= \frac{8.686}{T_k} \left(10^{-6}\right) \frac{dB}{\mu s} \\
&= 8.686 \left(10^{-6}\right) 2.81 \left(10^6\right)
\end{aligned}$$

$$\boxed{\therefore \quad L = 24.4 \ \ dB/\mu s}$$

Problem 6.3 Repeat Problem 6.2 but for backward volume spin waves at 5.4 GHz and $H_{DC} = 98.7 \, \text{kA/m}$ in the same film.

Solution 6.3

(a) Assuming $\Delta H = 60 \, \text{A/m}$ at 9.3 GHz, the Gilbert damping parameter is the same as in Problem 6.2

$$\alpha = \frac{\left|\frac{\gamma}{2\pi}\right| \mu_0 \Delta H}{2f}, \qquad \left|\frac{\gamma}{2\pi}\right| = 28 \ \ \text{GHz/T}$$

$$= \frac{(28 \times 10^9)(4\pi \times 10^{-7})60}{2(9.3 \times 10^9)}.$$

$$\boxed{\therefore \quad \alpha = 1.14 \left(10^{-4}\right)}$$

(b) The relaxation time is given by (6.21)

$$\frac{1}{T_k} = \alpha \frac{(\omega_0^2 + \omega^2)}{2\omega_0}$$

For backward volume waves

$$\omega_0 = |\gamma| \mu_0 H_{DC}$$
$$= 2\pi \left(28 \times 10^9\right) (4\pi \times 10^{-7})(98.7 \times 10^3)$$
$$= 2.18 \times 10^{10}$$

and

$$\omega = 2\pi \left(5.4 \times 10^9\right) = 3.39 \times 10^{10}$$

Hence,

$$\frac{1}{T_k} = \frac{1.14\left(10^{-4}\right)}{2} \left[2.18 + \frac{3.39^2}{2.18}\right] 10^{10} = 4.25 \times 10^6$$

The loss/time is given by (6.5)

$$L = \frac{8.686\left(10^{-6}\right)}{T_k} \text{ (dB/}\mu\text{s)}$$
$$= 8.686\,(4.25)$$

$$\boxed{\therefore \quad L = 36.9 \text{ (dB/}\mu\text{s)}}$$

6.2   Mode Orthogonality and Normalization

Consider two forward volume wave modes with wave numbers k_a and k_b propagating with the same frequency ω, having potential functions of the form[3]

$$\psi_{a,b}(\mathbf{r}) = \phi_{a,b}(y)e^{i\mathbf{k}_{a,b}\cdot\mathbf{r}}, \tag{6.22}$$

where $\mathbf{k}_{a,b}$ are vectors in the $x - y$ plane. From Walker's equation (5.13), the functions $\phi_{a,b}(z)$ satisfy

$$\frac{\partial^2 \phi_a}{\partial z^2} - k_a^2(1+\chi)\phi_a = 0, \tag{6.23a}$$

$$\frac{\partial^2 \phi_b}{\partial z^2} - k_b^2(1+\chi)\phi_b = 0. \tag{6.23b}$$

---

[3]This treatment follows Buris [3].

We can multiply (6.23a) by $\phi_b(z)$ and integrate by parts over z, as shown in [2, Sect. 6.3.1], to get

$$(k_a^2 - k_b^2) \int_{-\infty}^{\infty} (1 + \chi)\phi_a(z)\phi_b(z)dz = 0. \tag{6.24}$$

Since we started with the assumption of two modes, $a \neq b$, we obtain the orthogonality property of forward volume waves

$$\int_{-\infty}^{\infty} (1 + \chi)\phi_a(z)\phi_b(z)dz = 0, \quad a \neq b. \tag{6.25}$$

These orthogonal modes form a complete basis set. When we excite multiple thickness modes in the film, we can write the total potential as a linear combination of the orthogonal modes

$$\psi(\mathbf{r}) = \sum_{n=0}^{\infty} a_n \phi_n(z) e^{i\mathbf{k}_n \cdot \mathbf{r}}. \tag{6.26}$$

The total power carried by the waveguide becomes

$$P = \frac{1}{2}\text{Re} \int_{-\infty}^{\infty} (-i\omega\psi^*)\mathbf{b} \cdot \hat{\mathbf{k}}\,dz, \tag{6.27}$$

where P (mW/mm) is the power per unit length in the plane of the film perpendicular to the direction of propagation. For multiple forward volume wave modes, $\mathbf{b} \cdot \hat{\mathbf{k}}$ is given by

$$\mathbf{b} \cdot \hat{\mathbf{k}} = -i\mu_0 \sum_{m=0}^{\infty} a_m k_m (1 + \chi)\psi_m \tag{6.28}$$

which, along with (6.26) for the potential gives

$$P = -\frac{\omega\mu_0}{2}\text{Re}\left\{\sum_{m,n} a_n^* a_m k_m e^{i(\mathbf{k}_m - \mathbf{k}_n)\cdot\mathbf{r}} \int_{-\infty}^{\infty} (1 + \chi)\phi_n(z)\phi_m(z)dz\right\}. \tag{6.29}$$

$$= -\frac{\omega\mu_0}{2}\sum_n |a_n|^2 k_n \int_{-\infty}^{\infty} (1 + \chi)\phi_n(z)\phi_n(z)dz. \tag{6.30}$$

The total power is thus the sum of the powers carried in the individual modes.

Consider two ways to normalize the mode amplitudes ψ_0 (c.f. (6.36) and (6.42) in Problems 6.4 and 6.5, respectively).

(a) We can require that

$$-\int_{-\infty}^{\infty} (1+\chi)\phi_a(z)\phi_b(z)dz = \delta_{a,b}, \tag{6.31}$$

where $\delta_{a,b}$ is the Kronecker delta and is equal to 1 if $a = b$ and 0 otherwise. The coefficient ψ_0 is then

$$\psi_0 = \sqrt{\frac{2}{-(1+\chi)d}} \tag{6.32}$$

for all modes (even and odd).

(b) The mode functions can be normalized so that if $a_n = 1$ in (6.30), the power in the nth mode is P_n (mW/mm), and

$$\psi_{0n} = \sqrt{\frac{4P_n}{-(1+\chi)\omega\mu_0 k_n d}}. \tag{6.33}$$

Problem 6.4 Perform the integrations in

$$P = -\frac{\omega\mu_0}{2} \sum_n |a_n|^2 k_n \int_{-\infty}^{\infty} (1+\chi)\,\phi_n(z)\phi_n(z)dz, \tag{6.34}$$

and

$$-\int_{-\infty}^{\infty} (1+\chi)\,\phi_a(z)\phi_b(z)dz = \delta_{a,b}, \tag{6.35}$$

using the forward volume mode functions for the even and odd modes

$$\psi^{(e)}(\mathbf{r}) = \begin{cases} \psi_0 e^{k_t d/2} \cos(\sqrt{-(1+\chi)}\,k_t d/2)\, e^{i\mathbf{k}_t\cdot\mathbf{r}-k_t z}, & z > d/2 \\ \psi_0 \cos(\sqrt{-(1+\chi)}\,k_t z)\, e^{i\mathbf{k}_t\cdot\mathbf{r}}, & |z| \le d/2 \\ \psi_0 e^{k_t d/2} \cos(\sqrt{-(1+\chi)}\,k_t d/2)\, e^{i\mathbf{k}_t\cdot\mathbf{r}+k_t z}, & z < -d/2. \end{cases} \tag{6.36a}$$

$$\psi^{(o)}(\mathbf{r}) = \begin{cases} \psi_0 e^{k_t d/2} \sin(\sqrt{-(1+\chi)}\,k_t d/2)\, e^{i\mathbf{k}_t\cdot\mathbf{r}-k_t z}, & z > d/2 \\ \psi_0 \sin(\sqrt{-(1+\chi)}\,k_t z)\, e^{i\mathbf{k}_t\cdot\mathbf{r}}, & |z| \le d/2 \\ -\psi_0 e^{k_t d/2} \sin(\sqrt{-(1+\chi)}\,k_t d/2)\, e^{i\mathbf{k}_t\cdot\mathbf{r}+k_t z}, & z < -d/2. \end{cases} \tag{6.36b}$$

Starting with the mode normalization (6.35), verify that the normalization coefficient ψ_0 is given by

$$\psi_0 = \sqrt{\frac{2}{-(1+\chi)d}}, \tag{6.37}$$

for both even and odd modes. Similarly, starting with the power normalization (6.34) and taking $|a_n| = 1$, verify that the normalization coefficient for both even and odd modes in this case is given by

$$\psi_{0n} = \sqrt{\frac{4P_n}{-(1+\chi)\omega\mu_0 k_n d}}.$$

(6.38)

(Hint: some expressions can be simplified using the dispersion relation (5.43) from Chap. 5, along with the corresponding expression for odd modes (obtained from (5.44) with n odd)[4]:

$$\tan\left[\frac{k_t d}{2}\sqrt{-(1+\chi)}\right] = \frac{1}{\sqrt{-(1+\chi)}},$$

(6.39a)

$$-\cot\left[\frac{k_t d}{2}\sqrt{-(1+\chi)}\right] = \frac{1}{\sqrt{-(1+\chi)}},$$

(6.39b)

for even and odd modes, respectively.)

Solution 6.4

Let us define the integral

$$I = \int_{-\infty}^{\infty} (1+\chi)\phi_n(z)\phi_n(z)dz.$$

For **even modes,**

$$\phi_n^{(e)}(z) = \begin{cases} \psi_0 e^{k_n d/2}\cos(\frac{k_n d}{2}\sqrt{-(1+\chi)})e^{-k_n z}, & z > \frac{d}{2} \\ \psi_0 \cos(k_n z\sqrt{-(1+\chi)}), & |z| < \frac{d}{2} \\ \psi_0 e^{k_n d/2}\cos(\frac{k_n d}{2}\sqrt{-(1+\chi)})e^{k_n z}, & z < -\frac{d}{2} \end{cases}$$

Let us write the integral I in three parts

$$I = \int_{-\infty}^{-\frac{d}{2}} \phi_n \phi_n dz + \int_{-\frac{d}{2}}^{-\frac{d}{2}} (1+\chi)\phi_n \phi_n dz + \int_{\frac{d}{2}}^{\infty} \phi_n \phi_n dz$$

---

[4]Compare with (5.72) and (5.74) from Stancil and Prabhakar [2].

and define $\alpha \equiv k_n \sqrt{-(1+\chi)}$ such that

$$
I = \int_{-\infty}^{-\frac{d}{2}} dz \, \psi_0^2 e^{k_n d} \cos^2\left(\frac{\alpha d}{2}\right) e^{2k_n z}
$$

$$
+ \int_{-\frac{d}{2}}^{-\frac{d}{2}} (1+\chi)\psi_0^2 \cos^2(\alpha z) \, dz \int_{\frac{d}{2}}^{\infty} \psi_0^2 e^{k_n d} \cos^2\left(\frac{\alpha d}{2}\right) e^{-2k_n z} dz
$$

$$
= \psi_0^2 e^{k_n d} \cos^2\left(\frac{\alpha d}{2}\right) \left[\frac{e^{2k_n z}}{2k_n} \Big|_{-\infty}^{-d/2} - \frac{e^{-2k_n z}}{2k_n} \Big|_{d/2}^{\infty} \right]
$$

$$
+ \psi_0^2 (1+\chi) \left[\frac{\sin(\alpha z)\cos(\alpha z)}{2\alpha} + \frac{z}{2} \right]_{-d/2}^{d/2}
$$

$$
= \psi_0^2 e^{k_n d} \cos^2\left(\frac{\alpha d}{2}\right) \left[\frac{e^{-k_n d}}{k_n} \right] + \psi_0^2 (1+\chi) \left[\frac{\sin\left(\frac{\alpha d}{2}\right)\cos\left(\frac{\alpha d}{2}\right)}{\alpha} + \frac{d}{2} \right]
$$

$$
= \frac{\psi_0^2}{k_n} \cos^2\left(\frac{\alpha d}{2}\right) - \frac{\psi_0^2}{k_n} \sqrt{-(1+\chi)} \left[\sin\left(\frac{\alpha d}{2}\right) \cos\left(\frac{\alpha d}{2}\right) + \frac{\alpha d}{2} \right]
$$

But we also know (see (6.39a))

$$
\tan\left(\frac{\alpha d}{2}\right) = \frac{1}{\sqrt{-(1+\chi)}}
$$

or equivalently,

$$
\sqrt{-(1+\chi)} = \frac{\cos\left(\frac{\alpha d}{2}\right)}{\sin\left(\frac{\alpha d}{2}\right)}
$$

Hence

$$
I = \frac{\psi_0^2}{k_n} \cos^2\left(\frac{\alpha d}{2}\right) - \frac{\psi_0^2}{k_n} \cos^2\left(\frac{\alpha d}{2}\right) - \frac{\psi_0^2}{k_n} \sqrt{-(1+\chi)} \, k_n \sqrt{-(1+\chi)} \frac{d}{2}
$$
$$
= +\psi_0^2 (1+\chi)\frac{d}{2}
$$

If we assume the normalization (6.35), then it follows that $-I = 1$ and

$$
1 = -\psi_0^2 (1+\chi)\frac{d}{2}.
$$

Consequently,

$$
\boxed{\psi_0 = \sqrt{\frac{2}{-(1+\chi)d}}}
$$

If instead we choose the power normalization (6.34)

$$P = -\frac{\omega\mu_0}{2} \sum_n |a_n|^2 k_n \int_{-\infty}^{\infty} (1+\chi)\phi_n(z)\phi_n(z)dz$$

we obtain

$$P_n = -\frac{\omega\mu_0}{2} k_n I = -(1+\chi)\frac{\omega\mu_0}{2} k_n \psi_0^2 \frac{d}{2}.$$

We rearrange terms to find

$$\boxed{\psi_{0n} = \sqrt{\frac{4P_n}{-(1+\chi)\omega\mu_0 k_n d}}.} \tag{6.33}$$

For odd modes,

$$\phi_n^{(0)}(z) = \begin{cases} \psi_0 e^{\frac{k_n d}{2}} \sin\left(\frac{\alpha d}{2}\right) e^{-k_n z} & z > \frac{d}{2} \\ \psi_0 \sin(\alpha z) & |z| < \frac{d}{2} \\ \psi_0 e^{\frac{k_n d}{2}} \sin\left(\frac{\alpha d}{2}\right) e^{k_n z} & z < -\frac{d}{2} \end{cases}$$

$$I = \underbrace{\int_{-\infty}^{\frac{-d}{2}} \phi_n^2 dz + \int_{\frac{d}{2}}^{\infty} \phi_n^2 dz}_{\frac{\psi_0^2}{k_n}\sin^2\left(\frac{\alpha d}{2}\right) \text{ (is the same as before, but } \cos^2 \to \sin^2)} \qquad + \qquad \underbrace{(1+\chi)\int_{\frac{-d}{2}}^{\frac{d}{2}} \phi_n^2 dz}_{\psi_0^2(1+\chi)\left[\frac{-\sin\left(\frac{\alpha d}{2}\right)\cos\left(\frac{\alpha d}{2}\right)}{\alpha} + \frac{d}{2}\right]}$$

But for odd modes,

$$\cot\left(\frac{\alpha d}{2}\right) = \frac{-1}{\sqrt{-(1+\chi)}}$$

$$\Rightarrow \quad \sqrt{-(1+\chi)} = -\frac{\sin\left(\frac{\alpha d}{2}\right)}{\cos\left(\frac{\alpha d}{2}\right)}.$$

Therefore

$$I = \frac{\psi_0^2}{k_n}\sin^2\left(\frac{\alpha d}{2}\right) - \frac{\psi_0^2}{k_n}\sin^2\left(\frac{\alpha d}{2}\right) + \psi_0^2(1+\chi)\frac{d}{2}$$

$$= \psi_0^2(1+\chi)\frac{d}{2}$$

$$\boxed{\therefore \quad I = \psi_0^2(1+\chi)\frac{d}{2}.}$$

This is identical to the integral over even mode functions, so the normalizations are the same as before.

Problem 6.5 Perform the integrations in

$$P = -\frac{\omega\mu_0}{2}\sum_n |a_n|^2 k_n \int_{-\infty}^{\infty} \phi_n(y)\phi_n(y)\mathrm{d}y, \qquad (6.40)$$

and

$$\int_{-\infty}^{\infty} \phi_a(y)\phi_b(y)\mathrm{d}y = \delta_{a,b} \qquad (6.41)$$

using the backward volume mode functions

$$\psi^{(o)}(\mathbf{r}) = \begin{cases} \psi_0 e^{k_z d/2}\sin(k_y d/2)e^{i\nu k_z z - k_z y}, & y > d/2, \\ \psi_0 \sin(k_y y)e^{i\nu k_z z}, & |y| \leq d/2 \\ -\psi_0 e^{k_z d/2}\sin(k_y d/2)e^{i\nu k_z z + k_z y}, & y < -d/2 \end{cases} \qquad (6.42a)$$

$$\psi^{(e)}(\mathbf{r}) = \begin{cases} \psi_0 e^{k_z d/2}\cos(k_y d/2)e^{i\nu k_z z - k_z y}, & y > d/2, \\ \psi_0 \cos(k_y y)e^{i\nu k_z z}, & |y| \leq d/2 \\ \psi_0 e^{k_z d/2}\cos(k_y d/2)e^{i\nu k_z z + k_z y}, & y < -d/2. \end{cases} \qquad (6.42b)$$

For both even and odd modes, verify that the mode normalization coefficients are given by

$$\psi_0 = \sqrt{\frac{2}{d}}, \qquad (6.43)$$

$$\psi_{0n} = \sqrt{\frac{4|P_n|}{\omega\mu_0 k_n d}}, \qquad (6.44)$$

for the normalizations given by (6.41) and (6.40), respectively. (Hint: some expressions can be simplified using the dispersion relation (5.46a) re-written separately for odd and even values of n as follows:

$$\tan\left[\frac{k_z d}{2\sqrt{-(1+\chi)}}\right] = \sqrt{-(1+\chi)}, \qquad (6.45a)$$

$$-\cot\left[\frac{k_z d}{2\sqrt{-(1+\chi)}}\right] = \sqrt{-(1+\chi)}, \qquad (6.45b)$$

for odd and even modes, respectively. For further details, see [2, Sect. 5.5].)

Solution 6.5

Let us define the integral

$$I = \int_{-\infty}^{\infty} \phi_n(y)\phi_n(y)\mathrm{d}y$$

For **even modes,**

$$
\phi_n^{(e)}(y) = \begin{cases} \psi_0 e^{k_n d/2} \cos(\frac{k_n d}{2\sqrt{-(1+\chi)}}) e^{-k_n y} & y > \frac{d}{2} \\[2mm] \psi_0 \cos(\frac{k_n y}{\sqrt{-(1+\chi)}}) & |y| < \frac{d}{2} \\[2mm] \psi_0 e^{k_n d/2} \cos(\frac{k_n d}{2\sqrt{-(1+\chi)}}) e^{k_n y} & y < -\frac{d}{2} \end{cases}
$$

Let us write the integral in three parts

$$
I = \int_{-\infty}^{-\frac{d}{2}} \phi_n \phi_n dy + \int_{-\frac{d}{2}}^{-\frac{d}{2}} \phi_n \phi_n dy + \int_{\frac{d}{2}}^{\infty} \phi_n \phi_n dy
$$

and define $\alpha \equiv \dfrac{k_n}{\sqrt{-(1+\chi)}}$ such that

$$
I = \int_{-\infty}^{-\frac{d}{2}} \psi_0^2 e^{k_n d} \cos^2\left(\frac{\alpha d}{2}\right) e^{2k_n y} dy + \int_{-\frac{d}{2}}^{-\frac{d}{2}} \psi_0^2 \cos^2(\alpha y)\, dy
$$

$$
+ \int_{\frac{d}{2}}^{\infty} \psi_0^2 e^{k_n d} \cos^2\left(\frac{\alpha d}{2}\right) e^{-2k_n y} dy
$$

$$
= \psi_0^2 e^{k_n d} \cos^2\left(\frac{\alpha d}{2}\right) \left[\frac{e^{2k_n y}}{2k_n} \Big|_{-\infty}^{-d/2} - \frac{e^{-2k_n y}}{2k_n} \Big|_{d/2}^{\infty} \right]
$$

$$
+ \psi_0^2 \left[\frac{\sin(\alpha y)\cos(\alpha y)}{2\alpha} + \frac{y}{2} \right]_{-d/2}^{d/2}
$$

$$
= \psi_0^2 e^{k_n d} \cos^2\left(\frac{\alpha d}{2}\right) \left[\frac{e^{-k_n d}}{k_n} \right] + \psi_0^2 \left[\frac{\sin\left(\frac{\alpha d}{2}\right)\cos\left(\frac{\alpha d}{2}\right)}{\alpha} + \frac{d}{2} \right]
$$

$$
= \frac{\psi_0^2}{k_n} \cos^2\left(\frac{\alpha d}{2}\right) + \psi_0^2 \left[\sqrt{-(1+\chi)} \frac{\sin\left(\frac{\alpha d}{2}\right)\cos\left(\frac{\alpha d}{2}\right)}{k_n} + \frac{d}{2} \right]
$$

But we also know from (6.45b)

$$
-\cot\left(\frac{\alpha d}{2}\right) = \sqrt{-(1+\chi)}
$$

or equivalently,

$$
\sqrt{-(1+\chi)} = -\frac{\cos(\frac{\alpha d}{2})}{\sin(\frac{\alpha d}{2})}.
$$

Hence

$$
I = \frac{\psi_0^2}{k_n} \cos^2\left(\frac{\alpha d}{2}\right) - \frac{\psi_0^2}{k_n} \cos^2\left(\frac{\alpha d}{2}\right) + \psi_0^2 \frac{d}{2}
$$

$$
= +\psi_0^2 \frac{d}{2}
$$

If we assume the orthonormal condition (6.41)

$$\int_{-\infty}^{\infty} \phi_a(y)\phi_b(y)\mathrm{d}y = \delta_{a,b}$$

and set $a = b$, then it follows that $I = 1$ and

$$1 = \psi_0^2 \frac{d}{2}.$$

Consequently,

$$\boxed{\psi_0 = \sqrt{\frac{2}{d}}}$$

If instead we choose the power normalization (6.40)

$$P = -\frac{\omega\mu_0}{2}\sum_n |a_n|^2 k_n \int_{-\infty}^{\infty} \phi_n(y)\phi_n(y)\mathrm{d}y$$

we obtain

$$P_n = -\frac{\omega\mu_0}{2}k_n I = -\frac{\omega\mu_0}{2}k_n \psi_0^2 \frac{d}{2}.$$

In this case, the minus sign just gives the direction of energy flow, so using magnitudes we rearrange terms to find

$$\boxed{\psi_{0n} = \sqrt{\frac{4|P_n|}{\omega\mu_0 k_n d}}.}$$

For **odd modes,**

$$\phi_n^{(0)}(y) = \begin{cases} \psi_0 e^{\frac{k_n d}{2}} \sin\left(\frac{\alpha d}{2}\right) e^{-k_n y}, & y > \frac{d}{2} \\ \psi_0 \sin(\alpha y), & |y| < \frac{d}{2} \\ -\psi_0 e^{\frac{k_n d}{2}} \sin\left(\frac{\alpha d}{2}\right) e^{k_n y}, & y < -\frac{d}{2} \end{cases}$$

and,

$$I = \underbrace{\int_{-\infty}^{\frac{-d}{2}} \phi_n^2 \mathrm{d}z + \int_{\frac{d}{2}}^{\infty} \phi_n^2 \mathrm{d}z}_{\frac{\psi_0^2}{k_n}\sin^2\left(\frac{\alpha d}{2}\right)\text{ (is the same as before, but } \cos^2 \to \sin^2)} + \underbrace{\int_{\frac{-d}{2}}^{\frac{d}{2}} \phi_n^2 \mathrm{d}z}_{\psi_0^2\left[\frac{-\sin\left(\frac{\alpha d}{2}\right)\cos\left(\frac{\alpha d}{2}\right)}{\alpha} + \frac{d}{2}\right]}$$

From the dispersion relation (6.45a)

$$\tan\left(\frac{\alpha d}{2}\right) = \sqrt{-(1+\chi)}$$

$$\Rightarrow \quad \sqrt{-(1+\chi)} = \frac{\sin\left(\frac{\alpha d}{2}\right)}{\cos\left(\frac{\alpha d}{2}\right)}.$$

Therefore

$$I = \frac{\psi_0^2}{k_n}\sin^2\left(\frac{\alpha d}{2}\right) - \frac{\psi_0^2}{k_n}\sin^2\left(\frac{\alpha d}{2}\right) + \psi_0^2\frac{d}{2}$$

$$\therefore \quad I = \psi_0^2\frac{d}{2}.$$

This is identical to the integral over even mode functions, so the normalizations are the same as before.

Problem 6.6 The power per unit width carried by surface spin waves is given by (6.27)

$$P = \frac{1}{2}\mathrm{Re}\int_{-\infty}^{\infty}(-i\omega\psi^*)\mathbf{b}\cdot\hat{\mathbf{k}}\,dy. \tag{6.46}$$

Evaluate this integral using the mode functions

$$\psi_\nu(\mathbf{r}) = \begin{cases} \psi_0\left(e^{kd} + p(\nu)\right)e^{-ky+i\nu kx}, & y > d/2 \\ \psi_0\left(e^{ky} + p(\nu)e^{-ky}\right)e^{i\nu kx}, & |y| \leq d/2 \\ \psi_0\left(1 + p(\nu)e^{kd}\right)e^{ky+i\nu kx}, & y < -d/2 \end{cases} \tag{6.47}$$

and verify the normalization coefficient

$$\psi_{0\nu} = \sqrt{\frac{P}{-(1+\chi)\omega\mu_0 p(\nu)kd}}. \tag{6.48}$$

Here $p(\nu) \equiv \psi_{0-}/\psi_{0+}$ and is given equivalently by the following two expressions:

$$p(\nu) \equiv \frac{\psi_{0-}}{\psi_{0+}} = \frac{\chi - \nu\kappa}{\chi + 2 + \nu\kappa}e^{-kd}, \tag{6.49a}$$

$$p(\nu) = \frac{\chi + 2 - \nu\kappa}{\chi + \nu\kappa}e^{kd}. \tag{6.49b}$$

Remember that $\mathbf{h}$ has both y and x components. (Hint: some expressions can be simplified using the dispersion relation (5.48a))

$$e^{-2kd} = \frac{(\chi + 2)^2 - \kappa^2}{\chi^2 - \kappa^2},$$

and the definitions of the parameter p from (6.49). This calculation is rather tedious; be forewarned!)

Solution 6.6

From (6.47), the surface wave potential function is given by

$$
\psi_\nu =
\begin{cases}
\psi_0 \left(e^{kd} + p\left(\nu\right)\right) e^{-ky + i\nu kx}, & y > \frac{d}{2} \\[2mm]
\psi_0 \left(e^{ky} + p\left(\nu\right) e^{-ky}\right) e^{i\nu kx}, & |y| \le \frac{d}{2} \\[2mm]
\psi_0 \left(1 + p\left(\nu\right) e^{kd}\right) e^{ky + i\nu kx}, & y < -\frac{d}{2}
\end{cases}
$$

where

$$
p\left(\nu\right) = \frac{\chi - \nu k}{\chi - 2 + \nu k}\, e^{-kd} = \frac{\chi + 2 - \nu k}{\chi + \nu k}\, e^{kd}
$$

In the integral, $\hat{k} = \nu \hat{x}, \nu = \pm 1$. Therefore, we are interested in b_x in (6.46). Keeping in mind that there are both x and y components of $\bar{h}$, we have

$$
b_x = -\mu_0 \left[(1 + \chi) \frac{\partial \psi}{\partial x} - i\kappa \frac{\partial \psi}{\partial y} \right]
$$

or,

$$
b_x = -\mu_0
\begin{cases}
\frac{\partial \psi}{\partial x} & y > \frac{d}{2} \\[2mm]
(1 + \chi) \frac{\partial \psi}{\partial x} - i\kappa \frac{\partial \psi}{\partial y} & |y| \le \frac{d}{2} \\[2mm]
\frac{\partial \psi}{\partial x} & y < -\frac{d}{2}
\end{cases}
$$

It is convenient to divide the integral into 3 pieces

$$
P = \frac{1}{2} \mathrm{Re} \left\{ -i\omega\nu \int_{-\infty}^{\infty} \psi^* b_x \, dy \right\} = P_1 + P_2 + P_3
$$

where P_1 is over the region $y < -d/2$, P_2 is over the thickness of the film $|y| \le d/2$, and P_3 is over the region $y > d/2$. We consider each of these in turn

$$
\begin{aligned}
P_1 &= \frac{1}{2} \mathrm{Re} \left\{ -i\omega\nu \int_{-\infty}^{-d/2} \psi^* \left(-\mu_0 i\nu k \psi\right) dy \right\} \\[2mm]
&= \frac{1}{2} \mathrm{Re} \left\{ -\omega \mu_0 k \int_{-\infty}^{-d/2} \psi^* \psi \, dy \right\} \\[2mm]
&= \frac{-\omega \mu_0 k}{2} \int_{-\infty}^{-d/2} \psi^* \psi \, dy \\[2mm]
&= -\frac{\omega \mu_0 k}{2} \psi_0^2 \left(1 + p e^{kd}\right)^2 \int_{-\infty}^{-d/2} e^{2ky} \, dy \\[2mm]
&= -\frac{\omega \mu_0 k}{2} \psi_0^2 \left(1 + p e^{kd}\right)^2 \frac{e^{2ky}}{2k} \Big|_{-\infty}^{-d/2}
\end{aligned}
$$

$$
\boxed{\therefore \quad P_1 = \frac{-\omega \mu_0 \psi_0^2}{4} \left(1 + p e^{kd}\right)^2 e^{-kd}}
$$

P_3 has a very similar form

$$P_3 = -\frac{\omega\mu_0 k}{2} \int_{d/2}^{\infty} \psi^* \psi \, dy$$

$$= -\frac{\omega\mu_0 k}{2} \psi_0^2 \left(e^{kd} + p\right)^2 \int_{d/2}^{\infty} e^{-2ky} \, dy$$

Note

$$\int_{d/2}^{\infty} e^{-2ky} \, dy = -\int_{-d/2}^{-\infty} e^{2ky'} \, dy' = \int_{-\infty}^{-d/2} e^{2ky'} \, dy' = \frac{e^{-kd}}{2k}$$

Hence,

$$\boxed{P_3 = -\frac{\omega\mu_0 \psi_0^2}{4} \left(e^{kd} + p\right)^2 e^{-kd}}$$

Finally, the integral over the film is

$$P_2 = \frac{1}{2}\text{Re} \left\{-i\omega\nu \int_{-d/2}^{d/2} \psi^* \left(-\mu_0 (1+\chi) \frac{\partial\psi}{\partial x} + \mu_0 ik \frac{\partial\psi}{\partial y}\right) dy\right\}$$

Using $\dfrac{\partial\psi}{\partial x} = i\nu k\psi$, this can be written as

$$P_2 = -\underbrace{\frac{\omega\mu_0 k (1+\chi)}{2} \int_{-d/2}^{d/2} \psi^* \psi \, dy}_{I_1} + \underbrace{\frac{\omega\mu_0 \nu\kappa}{2} \int_{-d/2}^{d/2} \psi^* \frac{\partial\psi}{\partial y} dy}_{I_2}$$

Now

$$I_1 = \psi_0^2 \int_{-d/2}^{d/2} \left(e^{ky} + pe^{-ky}\right)^2 dy$$

$$= \psi_0^2 \int_{-d/2}^{d/2} \left[e^{2ky} + 2p + p^2 e^{-2ky}\right] dy$$

$$= \psi_0^2 \left[2pd + \int_{-d/2}^{d/2} e^{2ky} dy + p^2 \int_{-d/2}^{d/2} e^{-2ky} dy\right]$$

$$= \psi_0^2 \left[2pd + (1+p^2) \int_{-d/2}^{d/2} e^{2ky} dy\right],$$

since

$$\int_{-d/2}^{d/2} e^{2ky} dy = \int_{-d/2}^{d/2} e^{-2ky} dy.$$

Performing the integration gives

$$I_1 = \psi_0^2 \left[2pd + \left(1 + p^2\right) \frac{\left(e^{kd} - e^{-kd}\right)}{2k} \right].$$

Similarly

$$
\begin{aligned}
I_2 &= \psi_0^2 k \int_{-d/2}^{d/2} \left(e^{ky} + pe^{-ky}\right) \left(e^{ky} - pe^{-ky}\right) dy \\
&= \psi_0^2 k \int_{-d/2}^{d/2} \left(e^{2ky} - p^2 e^{-2ky}\right) dy \\
&= \psi_0^2 k \left(1 - p^2\right) \int_{-d/2}^{d/2} e^{2ky} dy \\
&= \psi_0^2 k \left(1 - p^2\right) \frac{\left(e^{kd} - e^{-kd}\right)}{2k} \\
\therefore \quad I_2 &= \psi_0^2 \frac{\left(1 - p^2\right)}{2} \left(e^{kd} - e^{-kd}\right)
\end{aligned}
$$

Putting I_1 and I_2 back into the expression for P_2

$$
\begin{aligned}
P_2 &= \frac{-\omega \mu_0 k \left(1 + \chi\right)}{2} \psi_0^2 \left[2pd + \left(1 + p^2\right) \frac{\left(e^{kd} - e^{-kd}\right)}{2k} \right] \\
&\quad + \frac{\omega \mu_0 \nu \kappa}{2} \psi_0^2 \frac{\left(1 - p^2\right)}{2} \left(e^{kd} - e^{-kd}\right) \\
&= -\psi_0^2 \omega \mu_0 \left(1 + \chi\right) pkd - \psi_0^2 \frac{\omega \mu_0 \left(1 + \chi\right)}{4} \left(1 + p^2\right) \left(e^{kd} - e^{-kd}\right) \\
&\quad + \psi_0^2 \frac{\omega \mu_0 \nu \kappa}{4} \left(1 - p^2\right) \left(e^{kd} - e^{-kd}\right)
\end{aligned}
$$

Adding all three pieces gives

$$P = P_1 + P_2 + P_3$$

i.e.,

$$
\begin{aligned}
\frac{4P}{\omega \mu_0 \psi_0^2} &= - \left(1 + pe^{kd}\right)^2 e^{-kd} - 4 \left(1 + \chi\right) pkd \\
&\quad - \left(1 + \chi\right) \left(1 + p^2\right) \left(e^{kd} - e^{-kd}\right) \\
&\quad + \nu \kappa \left(1 - p^2\right) \left(e^{kd} - e^{-kd}\right) - \left(e^{kd} + p\right)^2 e^{-kd}
\end{aligned}
$$

$$\therefore \frac{4P}{\omega\mu_0\psi_0^2} + 4\left(1+\chi\right)pkd$$

$$= -\left(e^{-kd} + 2p + p^2 e^{kd}\right) - \left(1+\chi\right)\left(1+p^2\right)\left(e^{kd} - e^{-kd}\right)$$
$$+ \nu\kappa\left(1 - p^2\right)\left(e^{kd} - e^{-kd}\right) - \left(e^{kd} + 2p + p^2 e^{-kd}\right)$$
$$= -4p - e^{-kd}\left[1 - \left(1+\chi\right)\left(1+p^2\right) + \nu\kappa\left(1 - p^2\right) + p^2\right]$$
$$- e^{kd}\left[p^2 + \left(1+\chi\right)\left(1+p^2\right) - \nu\kappa\left(1 - p^2\right) + 1\right]$$
$$= -4p - e^{-kd}\left[-\chi + \nu\kappa + p^2\left(-\chi - \nu\kappa\right)\right]$$
$$- e^{kd}\left[\chi + 2 - \nu\kappa + p^2\left(\chi + 2 + \nu\kappa\right)\right]$$

But

$$p^2 = \frac{(\chi - \nu\kappa)}{(\chi + 2 + \nu\kappa)}\frac{(\chi + 2 - \nu\kappa)}{(\chi + \nu\kappa)}$$

Therefore,

$$\frac{4P}{\omega\mu_0\psi_0^2} + 4\left(1+\chi\right)pkd$$

$$= -4p + e^{-kd}\left[\chi - \nu\kappa + \frac{(\chi - \nu\kappa)}{(\chi + 2 + \nu\kappa)}\frac{(\chi + 2 - \nu\kappa)}{(\chi + \nu\kappa)}(\chi + \nu\kappa)\right]$$
$$- e^{kd}\left[\chi + 2 - \nu\kappa + \frac{(\chi - \nu\kappa)}{(\chi + 2 + \nu\kappa)}\frac{(\chi + 2 - \nu\kappa)}{(\chi + \nu\kappa)}(\chi + 2 + \nu\kappa)\right]$$
$$= -4p + e^{-kd}\left(\chi - \nu\kappa\right)\left[1 + \frac{(\chi + 2 - \nu\kappa)}{(\chi + 2 + \nu\kappa)}\right]$$
$$- e^{kd}\left(\chi + 2 - \nu\kappa\right)\left[1 + \frac{\chi - \nu\kappa}{\chi + \nu\kappa}\right]$$
$$= -4p + e^{-kd}\frac{(\chi - \nu\kappa)}{(\chi + 2 + \nu\kappa)}\left[\chi + 2 + \nu\kappa + \chi + 2 - \nu\kappa\right]$$
$$- e^{kd}\frac{(\chi + 2 - \nu\kappa)}{(\chi + \nu\kappa)}\left[\chi + \nu\kappa + \chi - \nu\kappa\right]$$

But recall,

$$p = \frac{\chi - \nu\kappa}{\chi + 2 + \nu\kappa}e^{-kd} = \frac{\chi + 2 - \nu\kappa}{\chi + \nu\kappa}e^{kd}$$

Thus

$$\frac{4P}{\omega\mu_0\psi_0^2} + 4\left(1+\chi\right)pkd = -4p + 2p\left(\chi + 2\right) - p2\chi$$
$$= -4p + 2p\chi + 4p - 2p\chi$$
$$= 0,$$

and

$$\psi_0 = \psi_{0\nu} = \left[\frac{P}{-(1+\chi)\,\omega\mu_0 p(\nu)kd} \right]^{1/2}$$

6.3   Radiation Resistance

Current filaments are commonly used as transducers to excite spin waves in thin films.[5] The input terminals of the current filament present a complex impedance in general. The real part of this impedance represents the power that is radiated into the spin wave domain (neglecting ohmic resistance and other losses in the transducer) and is referred to as the *radiation resistance*. In this section, we consider expressions for the radiation resistance for the three principal spin wave modes in a thin film, and refer the reader to [2, Sect. 6.4.1] for treatment of the insertion and return losses experimentally observed in common excitation structures.

6.3.1   Volume Waves

In the case of volume waves, multiple modes are excited that propagate in both directions from the current filament. In general, the total radiation resistance from all the modes is

$$R_r = \sum_{m=0}^{\infty} R_{rm} = \sum_{m=0}^{\infty} r_{rm}l, \tag{6.50}$$

where r_{rm} is the radiation resistance per unit length for the mth mode, and l is the length of the current filament. It is often a good approximation to keep only the lowest-order mode in which case $R_r \approx R_{r0}$.

Mode orthogonality relations are useful in estimating the the power coupled from a microstrip antenna to the forward and backward volume waves. Consider the excitation by a current filament I, located at $(-\xi, s)$ above the surface of the film, as shown in Fig. 6.1. The current filament imposes a magnetoquasistatic potential, which we express as a series of orthogonal functions that include the volume wave thickness modes. The mode expansion can be written as

$$\psi_I(y = -\xi, z) = \sum_{n=0}^{\infty} g_n \phi_n(z)e^{jk_n\xi} + \sum_{n=0}^{\infty} d_n f_n(z), \tag{6.51}$$

where $\phi_n(z)$ are the volume mode functions, and $f_n(z)$ are any additional functions (orthogonal to the $\phi_n(z)$) needed to make the set complete. To find the potential ψ_I it is convenient to consider a shifted coordinate system with its origin coinciding

---

[5]Excitation of magnetostatic spin waves by a current has been discussed by a number of authors, e.g., [4–11].

Fig. 6.1 Current element, I, at a height s above the surface of a ferrite film of thickness d. We assume that the structure is uniform along the x direction. (Reproduced from [2] with permission.)

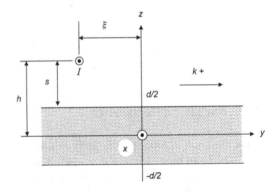

with the location of the current. In this new "primed" coordinate system the potential ψ_I is derived from Ampere's law for the field from a current carrying filament by equating these two forms

$$\mathbf{H}_I = \hat{\mathbf{e}}_{\phi'} \frac{1}{2\pi r'}, \tag{6.52}$$

$$\mathbf{H}_I = -\nabla \psi_I = -\hat{\mathbf{e}}_{\phi'} \frac{1}{r'} \frac{\partial \psi_I}{\partial \phi'} \tag{6.53}$$

to obtain

$$\psi_I = -\frac{I}{2\pi}\phi' + C, \tag{6.54}$$

where C is set to zero without loss of generality, r' is the cylindrical radial coordinate, and ϕ' is the counterclockwise angle from the y' axis. For $z' < 0$, $\phi' = -\pi/2$ while for $z' > 0$, $\phi' = +\pi/2$. Hence, in the original coordinate system we write

$$\psi_I(y = -\xi, z) = \frac{I}{4}\big[1 - 2u(z - h)\big] \tag{6.55}$$

where $u(z)$ is the unit step function.

6.3.2   Forward Volume Waves

We can now find the coefficients by multiplying both sides of (6.51) by $(1 + \chi)\phi_m$ and integrating to get

$$g_m = -e^{-jk_m\xi} \int_{-\infty}^{\infty} (1 + \chi)\phi_m(z)\psi_I(y = -\xi, z)\mathrm{d}z. \tag{6.56}$$

Carrying out this integration gives the following result for forward volume waves:

$$g_m = \frac{I}{2k_m} \sqrt{\frac{2}{-(1 + \chi)d}} \cos\left[\frac{k_m d}{2}\sqrt{-(1 + \chi)}\right] e^{-k_m s - jk_m\xi}. \tag{6.57}$$

If the excitation current is a finite-width strip instead of a filament, the total amplitude can be approximated by the superposition of N current filaments as

$$
\begin{aligned}
a_m &= \sum_{q=1}^{N} g_m(I_q, h_q, \xi_q) \\
&= \frac{I}{2k_m} \sqrt{\frac{2}{-(1+\chi)d}} \cos\left[\frac{k_m d}{2}\sqrt{-(1+\chi)}\right] \sum_{q=1}^{N} I_q e^{-k_m s_q - j k_m \xi_q}.
\end{aligned}
\tag{6.58}
$$

We can also write $I_q = K_q \Delta\xi$, where K_q is a surface current density (A/m). Taking the limit $\Delta\xi \to 0$, as $N \to \infty$ converts the sum into an integral and yields the array factor

$$
F = e^{-k_m s} \int_{\xi_1}^{\xi_2} K(\xi) e^{-j k_m \xi} d\xi.
\tag{6.59}
$$

If we consider a uniform current distribution described by $K(\xi) = I/w$ for $-w/2 \le \xi \le w/2$ and $K = 0$ elsewhere, the resulting array factor is

$$
F = I e^{-k_m s} \frac{\sin(k_m w/2)}{k_m w/2}.
\tag{6.60}
$$

However, a better approximation is to assume that the current density peaks at the edges of the conductor, and the corresponding array factor is [6]

$$
F = I e^{-k_m s} J_0(k_m w/2).
\tag{6.61}
$$

From (6.30), the power per unit length in the mth even mode propagating in the $+y$ direction is

$$
\begin{aligned}
P_m^+ &= \frac{1}{2}\omega\mu_0 k_m |a_m|^2 \\
&= \frac{\omega\mu_0}{4[-(1+\chi)]k_m d} \cos^2\left[\frac{k_m d}{2}\sqrt{-(1+\chi)}\right] |F|^2.
\end{aligned}
\tag{6.62}
$$

The corresponding radiation resistance per unit length is given by

$$
r_{rm}^+ = \frac{2P_m^+}{I^2}.
\tag{6.63}
$$

Forward volume waves are reciprocal, i.e., waves are launched in both the $+y$ and $-y$ directions. Hence, the total radiation resistance per unit length from the mth even mode is simply twice r_{rm}^+, i.e.

$$
r_{rm} = \frac{\omega\mu_0}{[-(1+\chi)]k_m d} \cos^2\left[\frac{k_m d}{2}\sqrt{-(1+\chi)}\right] \left|\frac{F}{I}\right|^2, \quad m = 0, 2, 4, \ldots .
\tag{6.64a}
$$

A similar calculation for the odd modes gives

$$r_{rm} = \frac{\omega\mu_0}{[-(1+\chi)]k_m d} \sin^2\left[\frac{k_m d}{2}\sqrt{-(1+\chi)}\right] \left|\frac{F}{I}\right|^2, \quad m = 1, 3, 5, \ldots. \quad (6.64b)$$

6.3.3   Backward Volume Waves

The orthogonality property for backward volume wave modes is

$$\int_{-\infty}^{\infty} \phi_a(y)\phi_b(y)\mathrm{d}y = 0, \quad\quad\quad (6.65)$$

where we have assumed potential functions of the form

$$\psi_{a,b}(\mathbf{r}) = \phi_{a,b}(y)e^{i\mathbf{k}_{a,b}\cdot\mathbf{r}}, \quad\quad\quad (6.22)$$

with $\mathbf{k}_{a,b}$ as vectors in the $x - y$ plane. An analysis similar to that of forward volume modes yields the radiation resistance per unit length for waves launched in the $+y$ direction

$$r_{rm}^+ = \frac{2P_m^+}{I^2}$$

$$= \frac{\omega\mu_0}{2k_m d} \sin^2\left[\frac{k_m d}{2\sqrt{-(1+\chi)}}\right] \left|\frac{F}{I}\right|^2. \quad\quad (6.66)$$

Backward volume waves are also reciprocal. Hence, we obtain the total radiation resistance per unit length of the mth odd mode by multiplying the previous expression by two

$$r_{rm} = \frac{\omega\mu_0}{k_m d} \sin^2\left[\frac{k_m d}{2\sqrt{-(1+\chi)}}\right] \left|\frac{F}{I}\right|^2, \quad m = 1, 3, 5, \ldots. \quad (6.67a)$$

For the even modes, a similar calculation using the potential functions in (6.42b) gives

$$r_{rm} = \frac{\omega\mu_0}{k_m d} \cos^2\left[\frac{k_m d}{2\sqrt{-(1+\chi)}}\right] \left|\frac{F}{I}\right|^2, \quad m = 2, 4, 6, \ldots. \quad (6.67b)$$

The normalized array factor $|F/I|$ that is present in (6.64) and (6.67) has nulls at the zeros of the Bessel function in (6.61). Consequently, these appear as dips in the plots of the radiation resistance, on a log scale, as shown in Fig. 6.2. (The radiation resistance plots in [2, Sect. 6.4] have assumed a uniform current density in the transducer with (6.60) as the array factor.)

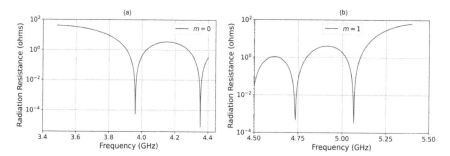

Fig. 6.2 Radiation resistance **a** with $H_{DC} = 239$ kA/m (3000 Oe) for $m = 0$ forward volume, and **b** with $H_{DC} = 98.7$ kA/m (1240 Oe) for $m = 1$ backward volume wave thickness modes. The other parameters, $M_S = 140$ kA/m ($4\pi M_S = 1760$ G), film thickness $d = 5$ μm, transducer width $w = 50$ μm, YIG-transducer spacing $s = 2$ μm, and transducer length $l = 3$ mm, with (6.61) for the array factor, are common to both plots. (Adapted from [2] with permission.)

6.3.4 Surface Waves

Surface waves exhibit field-displacement nonreciprocity and the radiation resistance is different for the two directions of propagation. We naturally expect that the mode localized at the surface nearest the transducer is the one that will be strongly excited. Thus, unlike volume waves, the radiation of surface waves will be unidirectional, and the efficiency of the excitation is determined by the orientation of $\mathbf{k}$ relative to $\mathbf{H}$. The radiation resistance per unit length for surface waves traveling in the νy direction (where $\nu = \pm 1$) is given by [4]

$$r_r^{(\nu)} = \frac{\mu_0 \omega}{2kd} \left[\frac{1 + \chi}{(1 + \nu\kappa)^2 - (1 + \chi)^2} \right] \left| \frac{F}{I} \right|^2. \tag{6.68}$$

With $\nu = \pm 1$ in the denominator, $r_r^{(+)} > r_r^{(-)}$, as shown in Fig. 6.3.

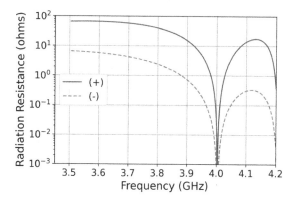

Fig. 6.3 Radiation resistances for both directions of surface wave propagation, with $H_{DC} = 51.7$ kA/m (650 Oe), $M_S = 140$ kA/m ($4\pi M_S = 1760$ G), film thickness $d = 5$ μm, transducer width $w = 50$ μm, YIG-transducer spacing $s = 2$ μm, and transducer length $l = 3$ mm, with (6.61) for the array factor. (Adapted from [2] with permission.)

Problem 6.7 Consider the radiation resistance for the lowest-order forward volume wave mode in the limit $k \to 0$.

(a) Using the dispersion relation (6.39), show that $kd[-(1+\chi)] \approx 2$ for small k.
(b) Using the result of part (a), take the limit of the radiation resistance (6.64a) as $k \to 0$ and show that

$$\lim_{k \to 0} r_{r0} = \frac{\omega \mu_0}{2}. \tag{6.69}$$

For simplicity, you may assume that the current is uniform in computing F/I.

Solution 6.7

(a) The dispersion relation for forward volume waves is given by (6.39):

$$\tan\left[\frac{kd}{2}\sqrt{-(1+\chi)}\right] = \frac{1}{\sqrt{-(1+\chi)}}$$

For small k, (provided $\lim_{k \to \infty} k\sqrt{-(1+\chi)} = 0$)

$$\tan\frac{kd}{2}\sqrt{-(1+\chi)} \approx \frac{kd}{2}\sqrt{-(1+\chi)}$$

Therefore

$$\frac{kd}{2}\sqrt{-(1+\chi)} \approx \frac{1}{\sqrt{-(1+\chi)}}$$

$$\boxed{kd\left[-(1+\chi)\right] \approx 2}$$

(b) Starting with (6.64a) and using (6.60) for F gives

$$r_{r0} = \frac{\omega \mu_0}{[-(1+\chi)]kd}\cos^2\left[\frac{kd}{2}\sqrt{-(1+\chi)}\right]e^{-2ks}\left[\frac{\sin\left(\frac{kw}{2}\right)}{\left(\frac{kw}{2}\right)}\right]^2$$

$$\lim_{k \to 0} r_{r0} = \lim_{k \to 0}\frac{\omega \mu_0}{[-(1+\chi)]kd}\underbrace{\cos^2\left[\frac{kd}{2}\sqrt{-(1+\chi)}\right]e^{-2ks}\left[\frac{\sin\left(\frac{kw}{2}\right)}{\left(\frac{kw}{2}\right)}\right]^2}_{\text{All tend to 1}})$$

$$= \lim_{k \to 0}\frac{\omega \mu_0}{[-(1+\chi)]kd}$$

Using the result from (a),

$$\boxed{\lim_{k \to 0} r_{r0} = \frac{\omega \mu_0}{2}}$$

Problem 6.8 Consider the radiation resistance for the lowest-order backward volume wave mode in the limit $k \to 0$.

(a) Using the dispersion relation (6.45a), show that $kd/[-(1+\chi)] \approx 2$ for small k.
(b) Using the result of part (a), take the limit of the radiation resistance (6.67a) as $k \to 0$ and show that the result is the same as that for forward volume waves; i.e.,

$$\lim_{k \to 0} r_{r1} = \frac{\omega \mu_0}{2}. \tag{6.70}$$

Solution 6.8

The dispersion relation (6.45a) for the lowest-order backward volume wave mode is

$$\tan \left[\frac{k_z d}{2\sqrt{-(1+\chi)}} \right] = \sqrt{-(1+\chi)}.$$

(a) For $k_z d / \sqrt{-(1+\chi)} \ll 1$, we can use the approximation $\tan \theta \approx \theta$ and write the dispersion relation as

$$\frac{k_z d}{2\sqrt{-(1+\chi)}} = \sqrt{-(1+\chi)}$$

or,

$$\boxed{\frac{k_z d}{-(1+\chi)} \approx 2}$$

(b) The radiation resistance for odd BVWs is given by

$$r_{rm} = \frac{\omega \mu_0}{k_m d} \sin \left[\frac{k_m d}{2\sqrt{-(1+\chi)}} \right]^2 \left| \frac{F}{I} \right|^2 \qquad m = 1, 3, 5 \ldots$$

The lowest-order backward volume wave mode is given by $m = 1$. For $k_m = k_z$ and $k_z s, k_z w \ll 1$, the factor $|F/I|$ becomes

$$F \approx \int_{-w/2}^{w/2} K(\xi) d\xi = I.$$

Consequently, the factor $|F/I| \to 1$. (Note that we specifically showed this was true for the uniform current distribution used in Problem 6.7.) Making this substitution along with the approximation $\sin \theta \approx \theta$ gives

$$\lim_{k \to 0} r_{r1} = \frac{\omega \mu_0}{k_z d} \left[\frac{k_z d}{2\sqrt{-(1+\chi)}} \right]^2$$

$$= \omega \mu_0 \left[\frac{k_z d}{4(-(1+\chi))} \right]$$

Simplifying this expression using the result from Part (a) gives the desired result

$$\lim_{k_z \to 0} r_{r1} = \frac{\omega \mu_0}{2}$$

Problem 6.9 In this problem, we find the small-k limit for the surface wave radiation resistance per unit length.

(a) Expand the dispersion relation (5.48b) for small k and show that

$$kd \approx 2 \left[\omega^2 - \omega_0(\omega_0 + \omega_M) \right] / \omega_M^2. \tag{6.71}$$

(b) Using the result of part (a) along with the fact that $\omega^2 = \omega_0(\omega_0 + \omega_M)$ at the $k = 0$ band edge, take the limit of the radiation resistance (6.68) as $k \to 0$ and show that the result can be expressed as

$$\lim_{k \to 0} r_r^{(\nu)} = \frac{\omega \mu_0}{4} \frac{\omega_0 \omega_M}{[\omega_0 - \nu \omega]^2}. \tag{6.72}$$

Solution 6.9

(a) The dispersion relation (5.48b) is

$$\omega^2 = \omega_0(\omega_0 + \omega_M) + \frac{\omega_M^2}{4} \left[1 - e^{-2kd} \right].$$

Expanding the exponential for small k gives

$$\omega^2 \approx \omega_0(\omega_0 + \omega_M) + \frac{\omega_M^2}{2} kd.$$

Solving for kd, we obtain

$$kd \approx \frac{2}{\omega_M^2} \left(\omega^2 - \omega_0 (\omega_0 + \omega_M) \right).$$

(b) We are given (6.68)

$$r_{\mathrm{r}}^{(\nu)} = \frac{\mu_0\omega}{2kd}\left(\frac{1+\chi}{(1+\nu\kappa)^2 - (1+\chi)^2}\right)\left|\frac{F}{I}\right|^2$$

Using the definitions of χ and κ from (3.63a) and (3.63b), respectively, and the result from part(a), we have

$$1+\chi = \frac{\omega_0^2 + \omega_0\omega_{\mathrm{M}} - \omega^2}{\omega_0^2 - \omega^2} = \frac{-\omega_{\mathrm{M}}^2 kd}{2}\frac{1}{\omega_0^2 - \omega^2},$$

$$1+\nu\kappa = \frac{\omega_0^2 + \nu\omega\omega_{\mathrm{M}} - \omega^2}{\omega_0^2 - \omega^2}.$$

We also know that (see solution to Problem 6.8)

$$\lim_{k\to 0}\left|\frac{F}{I}\right|^2 = 1.$$

and that, $1+\chi \to 0$ as $k \to 0$. Substituting into (6.68) gives

$$r_{\mathrm{r}}^{(\nu)} = \frac{\mu_0\omega}{2kd}\left[\frac{1+\chi}{(1+\nu\kappa)^2 - (1+\chi)^2}\right]\left|\frac{F}{I}\right|^2.$$

$$= \frac{\mu_0\omega}{2\,\cancel{kd}}\left(\frac{-\omega_{\mathrm{M}}^2\,\cancel{kd}}{2}\frac{1}{\omega_0^2 - \omega^2}\right)\left[\frac{1}{(1+\nu\kappa)^2 - \cancel{(1+\chi)^2}}\right]$$

$$= \frac{\mu_0\omega}{4}\left(\frac{-\omega_{\mathrm{M}}^2}{\omega_0^2 - \omega^2}\right)\left(\frac{\omega_0^2 - \omega^2}{\omega_0^2 + \nu\omega\omega_{\mathrm{M}} - \omega^2}\right)^2.$$

Now, make the substitution $\omega_0^2 = \omega^2 - \omega_0\omega_{\mathrm{M}}$, and simplify to get

$$\boxed{\lim_{k\to 0} r_{\mathrm{r}}^{(\nu)} \approx \frac{\omega\mu_0}{4}\frac{\omega_0\omega_{\mathrm{M}}}{(\omega_0 - \nu\omega)^2}.}$$

References

1. D.D. Stancil, Phenomenological propagation loss theory for the magnetostatic waves in thin ferrite films. J. Appl. Phys. **59**, 218 (1986)
2. D.D. Stancil, A. Prabhakar, *Spin Waves: Theory and Applications* (Springer, New York, 2009)
3. N.E. Buris, *Magnetostatic wave propagation in inhomogeneous and anisotropic ferrite thin films*, Ph.D. dissertation (North Carolina State University, 1986)
4. P.R. Emtage, Interaction of magnetostatic waves with a current. J. Appl. Phys. **49**, 4475 (1978)
5. A.K. Ganguly, D.C. Webb, Microstrip excitation of magnetostatic surface waves, theory and experiment, in *IEEE Microwave Theory Technology*, vol. MTT-23 (1975), p. 998

6. A.K. Ganguly, D.C. Webb, C. Banks, Complex radiation impedance of microstrip-excited magnetostatic-surface waves. IEEE Trans. Microwave Theory Tech. **26**, 444 (1978)
7. B.A. Kalinikos, V.F. Dmitriev, Self-consistent calculation of radiation resistance of microstrip transducer of spin waves in a perpendicularly magnetized ferromagnetic film. Zh. Tech. Fiz **58**, 248 (1988)
8. B.A. Kalinikos, V.F. Dmitriev, Self-consistent theory of excitation of spin waves by microstrip transducer in tangentially magnetized layered structure. Radiotekhnica and Electronica **33**, 2248 (1988)
9. J.P. Parekh, Theory of magnetostatic forward volume wave excitation. J. Appl. Phys. **50**, 2452 (1979)
10. J.C. Sethares, Magnetostatic surface wave transducers. IEEE Trans. Microwave Theory Tech. **27**, 902 (1979)
11. J.C. Sethares, I.J. Weinberg, Theory of MSW transducers. Circuits Syst. Signal Process **4**, 41 (1985)

Variational Formulation for Magnetostatic Modes

7

The magnetostatic modes that we discussed in Chap. 5 were for geometries charac-terized by simple boundary shapes, uniform bias fields, and uniform materials. This was, however, an idealization, and in many cases, we must include the effects of material and field nonuniformities. Such effects are often used to control dispersion or to guide and localize the magnetostatic mode energy. We may also need to assess the effects of undesired inhomogeneities. The classical boundary value techniques used in Chap. 5 are not easily applicable in such cases, but variational formulations enable the treatment of arbitrary inhomogeneities in a relatively simple and elegant way. Consequently, we adopt a variational approach, based on the treatment in [1–5]. Tsutsumi et al. [6] and Sawado and Chang [7] have described other variational formulations for dipolar spin waves in thin films.

Problems 7.1–7.3 explore the relationship between the magnetostatic Lagrangian density, $\mathbf{b} \cdot \mathbf{h}^*$, and Laplace's and Walker's equations, while Problems 7.4–7.6 fill in derivations discussed in the text. In Problem 7.5, the form of the dispersion energy density expressed completely in terms of the small-signal magnetization is the basis for the effective quasi-particle number density. Problem 7.7 confirms that the exact dispersion relations for forward volume spin waves reduce to the large-k limit obtained from the variational approximation.

7.1 Calculus of Variations

In the Lagrangian formulation of classical mechanics, the equations of motion are obtained from the difference between the kinetic and potential energies of the system by finding a solution that is stationary with respect to small perturbations, i.e., where errors of order ϵ in the approximation for the solution function result in errors in the Lagrangian of order ϵ^2. For excitations in continuous media, the Lagrangian

© The Author(s), under exclusive license to Springer Nature Switzerland AG 2021
D. D. Stancil and A. Prabhakar, *Spin Waves*,
https://doi.org/10.1007/978-3-030-68582-9_7

can be obtained from a *Lagrangian density* $\mathcal{L}$ by integrating over the volume of interest. The Lagrangian is taken to be a *functional*—i.e., a quantity that depends on a *function*. In general, the functional depends on the unknown function, the gradient of the unknown function, and position:

$$L = \int_V \mathcal{L}(\psi, \nabla\psi, \mathbf{r}) dv. \tag{7.1}$$

The correct solution, i.e., the function ψ that makes L stationary, satisfies the *Euler–Lagrange* equation

$$\frac{\partial \mathcal{L}}{\partial \psi} - \nabla \cdot \frac{\partial \mathcal{L}}{\partial (\nabla\psi)} = 0, \tag{7.2}$$

subject to the natural boundary condition on the surface enclosing the volume

$$\hat{\mathbf{n}} \cdot \frac{\partial \mathcal{L}}{\partial (\nabla\psi)}\bigg|_S = 0. \tag{7.3}$$

In magnetostatic problems, the Lagrangian density can be taken to be

$$\mathcal{L} = \mathbf{b} \cdot \mathbf{h}^*, \tag{7.4}$$

where $\mathbf{h} = -\nabla\psi$, $\mathbf{b} = \overline{\mu} \cdot \mathbf{h}$, and $\overline{\mu}$ is the magnetic permeability given by (4.13).

Problem 7.1 Show that the function ϕ that minimizes the functional

$$I[\phi] = \frac{1}{2} \int_V (\nabla\phi)^2 \, dv \tag{7.5}$$

also satisfies Laplace's equation, $\nabla^2\phi = 0$.

Solution 7.1

To identify the Lagrangian density, we write

$$I[\phi] = \frac{1}{2} \int_V (\nabla\phi)^2 \, dv = \int_V \mathcal{L} \, dv,$$

from which we identify $\mathcal{L} = \frac{1}{2}(\nabla\phi)^2$. The Euler–Lagrange equation (7.2) is

$$\frac{\partial \mathcal{L}}{\partial \phi} - \nabla \cdot \frac{\partial \mathcal{L}}{\partial (\nabla\phi)} = 0$$

which becomes

$$-\nabla \cdot (\nabla\phi) = 0,$$

Fig. 7.1 Forward volume wave geometry. (Reproduced from [9] with permission.)

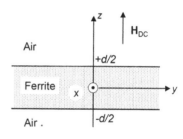

or

$$\boxed{\nabla^2 \phi = 0.}$$

Thus, we conclude that ϕ satisfies Laplace's equation. Note that in a medium where the permeability is a scalar constant and the potential is of the form $\psi = \phi \exp(i\alpha)$, where ϕ is a real function of position and α is independent of position, the Lagrangian density in (7.4) is the same as the one in this problem to within a numerical factor. We thus conclude that the magnetostatic potential must satisfy Laplace's equation in media with constant, scalar permeabilities.

Problem 7.2 According to the method of *Lagrange multipliers*, minimizing the functional $I[\psi(x)]$, subject to the constraint $J[\psi(x)] = \text{constant}$, is equivalent to minimizing the composite functional $I - \lambda J$ where λ is called the Lagrange multiplier [8, Chap. 12].

(a) Consider forward volume waves propagating in an infinite film as shown in Fig. 7.1 In this geometry, the only component of the small-signal field $\mathbf{h}$ that is transverse to the direction of propagation is $\mathbf{h} \cdot \hat{\mathbf{z}} = -\frac{\partial \psi}{\partial z} \equiv -\psi_z$. Show that minimizing the functional

$$I[\psi] = \int_{-\infty}^{\infty} \psi_z^* \psi_z \, dz \qquad (7.6)$$

is the same as requiring ψ to satisfy the one-dimensional Laplace equation $\partial^2 \psi / \partial z^2 = 0$.

(b) Consider the constraint

$$J = \int_{-\infty}^{\infty} [-(1 + \chi)] \psi^* \psi \, dz = \text{constant}, \qquad (7.7)$$

which is simply a normalization condition derived from the orthogonality relation (6.31) applied to a single mode. Find the Lagrange multiplier λ that makes the composite functional $\mu_0 (I - \lambda J)$ identical with

$$L = \int_V \mathbf{b} \cdot \mathbf{h}^* dv \qquad (7.8)$$

specialized to the present geometry. This leads us to the alternate interpretation of the *variational formulation* for forward volume waves: *the transverse variations of the potential must satisfy Laplace's equation subject to the normalization constraint (7.7)*.

Solution 7.2

(a) From the functional

$$I[\psi] = \int\limits_{-\infty}^{\infty} \psi_z^* \psi_z dz,$$

we obtain the Lagrangian density $\mathcal{L} = \psi_z^* \psi_z$. The Euler–Lagrange equation

$$\frac{\partial \mathcal{L}}{\partial \psi} - \nabla \cdot \frac{\partial \mathcal{L}}{\partial (\nabla \psi)} = 0$$

becomes

$$-\frac{\partial}{\partial z} \frac{\partial \mathcal{L}}{\partial \psi_z} = 0,$$

or

$$\boxed{\frac{\partial^2 \psi^*}{\partial z^2} = \frac{\partial^2 \psi}{\partial z^2} = 0.}$$

Thus, we conclude that ψ satisfies the one-dimensional Laplace equation.

You may recall that the Euler–Lagrange equation was obtained from the consideration of the Lagrangian density over a finite volume with specific natural boundary conditions on the surface. However, the geometry in Fig. 7.1 is open instead of a finite volume enclosed by a surface with specific boundary conditions, so you might wonder if the use of the Euler–Lagrange equation is justified. However, because the modes are bound to the magnetic film, the fields vanish as the surface recedes to infinity, thus satisfying the natural boundary condition (7.3).

(b) From (7.6) and (7.7) we have

$$\mu_0 (I - \lambda J) = \mu_0 \int_{-\infty}^{\infty} \psi_z^* \psi_z dz - \mu_0 \lambda \int_{-\infty}^{\infty} [-(1 + \chi)]\psi^* \psi \, dz. \qquad (7.9)$$

Our task is to find λ so that this is equivalent to $\int_V \overline{b} \cdot \overline{h}^* \, dv$ for this geometry. Using the definitions $\overline{h} = -\nabla \psi$, $\overline{b} = \overline{\overline{\mu}} \cdot \overline{h}$, and the permeability $\overline{\overline{\mu}}$ from (4.13),

$$\frac{1}{\mu_0} \int_V \overline{b} \cdot \overline{h}^* \, dv = \int_V dv \left\{ (1 + \chi) \left[\psi_x^* \psi_x + \psi_y^* \psi_y \right] \right.$$
$$\left. + \psi_z^* \psi_z - i\kappa \left[\psi_x^* \psi_y - \psi_y^* \psi_x \right] \right\}. \qquad (7.10)$$

Fig. 7.2 General backward
volume wave geometry.
(Reproduced from [9] with
permission.)

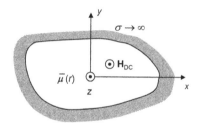

For the forward volume spin wave geometry, $\psi_x = ik_x\psi$, and $\psi_y = ik_y\psi$. Therefore

$$\frac{1}{\mu_0}\int_V \bar{b}\cdot\bar{h}^* \, dv = \int_V dv\left\{(1+\chi)\left[k_x^2 + k_y^2\right]\psi^*\psi\right.$$

$$\left. + \psi_z^*\psi_z - i\kappa k_x k_y \underbrace{\left[\psi^*\psi - \psi^*\psi\right]}_{0}\right\}.$$

Noting that the integrand does not depend on x or y (removed by the quadratic form $\psi^*\psi$), the integrations over x and y can be omitted. We thus have

$$\int_V \bar{b}\cdot\bar{h}^* \, dv \rightarrow \mu_0 \int_{-\infty}^{\infty} dz \left\{k_t^2 (1+\chi)\,\psi^*\psi + \psi_z^*\psi_z\right\}, \qquad (7.11)$$

where $k_t^2 = k_x^2 + k_y^2$. A comparison of (7.9) with (7.11) shows that they are the same if

$$\boxed{\lambda = k_t^2.}$$

We use the opportunity to also observe that setting (7.11) to zero, and solving for k_t^2, gives us a stationary expression for the forward volume wave dispersion relation,

$$k_t^2 = \frac{\int \psi_z^*\psi_z dz}{\int [-(1+\chi)]\psi^*\psi dz}. \qquad (7.31)$$

Problem 7.3 This problem applies the Lagrange multiplier analysis of Problem 7.2 to the case of backward volume waves in the geometry of Fig. 7.2.

(a) Show that minimizing the functional

$$I[\psi] = \int_S \left\{[-(1+\chi)]\left[\psi_x^*\psi_x + \psi_y^*\psi_y\right] + i\kappa\left[\psi_x^*\psi_y - \psi_y^*\psi_x\right]\right\} ds \qquad (7.12)$$

is the same as requiring ψ to satisfy the two-dimensional Walker equation

$$\frac{\partial}{\partial x}[(1+\chi)\psi_x] + \frac{\partial}{\partial y}[(1+\chi)\psi_y] - i\frac{\partial}{\partial x}[\kappa\psi_y] + i\frac{\partial}{\partial y}[\kappa\psi_x] = 0. \qquad (7.13)$$

(b) Consider the constraint

$$J = \int_S \psi^* \psi \, ds = \text{constant}, \qquad (7.14)$$

which is simply a normalization condition derived from the orthogonality relation

$$\int_{-\infty}^{\infty} \phi_a(y)\phi_b(y)dy = \delta_{a,b} \qquad (7.15)$$

applied to a single mode. Find the Lagrange multiplier λ that makes the composite functional $\mu_0(I - \lambda J)$ equivalent to L from (7.8) specialized to the present geometry. This leads us to the alternate interpretation of the variational formulation for backward volume waves: *the transverse variations of the potential must satisfy the two-dimensional Walker equation (7.13) subject to the normalization constraint (7.14).*

Solution 7.3

(a) Referring to (7.12), the Lagrangian density is

$$\mathcal{L} = -(1 + \chi)\left[\psi_x^* \psi_x + \psi_y^* \psi_y\right] + i\kappa\left[\psi_x^* \psi_y - \psi_y^* \psi_x\right].$$

This must satisfy the Euler–Lagrange equation

$$\frac{\partial \mathcal{L}}{\partial \psi} - \nabla \cdot \frac{\partial \mathcal{L}}{\partial (\nabla \psi)} = 0.$$

The Lagrangian density does not depend explicitly on ψ, so the Euler–Lagrange equation becomes

$$-\frac{\partial}{\partial x}\frac{\partial \mathcal{L}}{\partial \psi_x} - \frac{\partial}{\partial y}\frac{\partial \mathcal{L}}{\partial \psi_y} = 0$$

$$-\frac{\partial}{\partial x}\left[-(1+\chi)\psi_x^* - i\kappa\psi_y^*\right] - \frac{\partial}{\partial y}\left[-(1+\chi)\psi_y^* + i\kappa\psi_x^*\right] = 0$$

$$\frac{\partial}{\partial x}\left[(1+\chi)\psi_x^*\right] + \frac{\partial}{\partial y}\left[(1+\chi)\psi_y^*\right] + i\frac{\partial}{\partial x}\left[\kappa\psi_y^*\right] - i\frac{\partial}{\partial y}\left[\kappa\psi_x^*\right] = 0.$$

Taking the complex conjugate gives the general 2-D Walker's equation

$$\boxed{\frac{\partial}{\partial x}\left[(1+\chi)\psi_x\right] + \frac{\partial}{\partial y}\left[(1+\chi)\psi_y\right] - i\frac{\partial}{\partial x}\left[\kappa\psi_y\right] + i\frac{\partial}{\partial y}\left[\kappa\psi_x\right] = 0} \quad (7.16)$$

We also require for arbitrary $\delta\psi$

$$\frac{\partial \mathcal{L}}{\partial \psi^*} - \nabla \cdot \frac{\partial \mathcal{L}}{\partial (\nabla \psi^*)} = 0.$$

Repeating the calculation again leads to (7.16).

(b) From (7.12) and (7.14), we have

$$\mu_0 (I - \lambda J) = \mu_0 \int_S ds \left\{ -(1+\chi)\left[\psi_x^* \psi_x + \psi_y^* \psi_y \right] \right.$$
$$\left. + i\kappa \left[\psi_x^* \psi_y - \psi_y^* \psi_x \right] - \lambda \psi^* \psi \right\}. \tag{7.17}$$

Our task is to find λ so that this is equivalent to $\int_{-\infty}^{\infty} \overline{b} \cdot \overline{h}^* \, dv$ for this geometry. From (7.10), we have

$$\frac{1}{\mu_0} \int_V \overline{b} \cdot \overline{h}^* \, dv = \int_V dv \left\{ (1+\chi)\left[\psi_x^* \psi_x + \psi_y^* \psi_y \right] \right.$$
$$\left. + \psi_z^* \psi_z - i\kappa \left[\psi_x^* \psi_y - \psi_y^* \psi_x \right] \right\}.$$

For the backward volume spin wave geometry of Fig. 7.2, $\psi_z = i k_z \psi$. Therefore

$$\frac{1}{\mu_0} \int_V \overline{b} \cdot \overline{h}^* \, dv = \int_V dv \left\{ (1+\chi)\left[\psi_x^* \psi_x + \psi_y^* \psi_y \right] \right.$$
$$\left. + k_z^2 \psi^* \psi - i\kappa \left[\psi_x^* \psi_y - \psi_y^* \psi_x \right] \right\}.$$

Noting that the integrand does not depend on z (removed by the quadratic form $\psi^*\psi$), the volume integration can be replaced with a surface integration over the cross section. We then have

$$\int_V \overline{b} \cdot \overline{h}^* \, dv \rightarrow \mu_0 \int_S ds \left\{ (1+\chi)\left[\psi_x^* \psi_x + \psi_y^* \psi_y \right] \right.$$
$$\left. + k_z^2 \psi^* \psi - i\kappa \left[\psi_x^* \psi_y - \psi_y^* \psi_x \right] \right\}. \tag{7.18}$$

A comparison of (7.17) and (7.18) shows that $\mu_0 (I - \lambda J)$ corresponds to the negative of $\int_V \overline{b} \cdot \overline{h}^* \, dv$ if $\boxed{\lambda = k_z^2.}$ The expressions are, therefore, equivalent when

$$\boxed{\int_V \overline{b} \cdot \overline{h}^* \, dv = \mu_0 (I - \lambda J) = 0.}$$

Note that setting (7.18) to zero and solving for k_z^2 gives the stationary expression for the backward volume wave dispersion relation

$$k_z^2 = \int_S \frac{\left\{[-(1+\chi)][\psi_x^*\psi_x + \psi_y^*\psi_y] + i\kappa[\psi_x^*\psi_y - \psi_y^*\psi_x]\right\} ds}{\int_S \psi^*\psi \, ds}.$$

Problem 7.4 The instantaneous Zeeman energy density is given by

$$W_z(t) = -\mu_0 H_0 M_0 \cos\theta(t), \tag{7.19}$$

where $\theta(t)$ is the time-varying angle between the magnetization vector and the applied bias field.

(a) Expand $W_z(t)$ for small $\theta(t)$ and show that $W_z(t) = W_0 + w_z(t)$, where W_0 and $w_z(t)$ are static and time-varying components, respectively.
(b) Show that the time average of $w_z(t)$ from part (a) can be expressed as

$$\langle w_z(t) \rangle = \frac{\mu_0}{4} \frac{H_0}{M_0} |\mathbf{m}|^2. \tag{7.20}$$

Solution 7.4

(a) Expanding the cosine in (7.19) for for small $\theta(t)$

$$W_z(t) \approx -\mu_0 H_0 M_0 \left(1 - \frac{\theta^2}{2}\right). \tag{7.21}$$

This is of the form $W_z(t) = W_0 + w_z(t)$, where

$$\boxed{W_0 = -\mu_0 H_0 M_0, \quad w_z(t) = \frac{1}{2}\mu_0 H_0 M_0 \theta^2(t).}$$

(b) For small precession angles,

$$M_z = M_0 \cos\theta \approx M_0, \qquad m = M_0 \sin\theta \approx M_0\theta.$$

Also, for sinusoidally varying quantities,

$$\langle m^2(t) \rangle = \frac{1}{2}\text{Re}\left\{\overline{m} \cdot \overline{m}^*\right\} = \frac{1}{2}|\overline{m}|^2.$$

We can, therefore, write

$$w_z(t) = \frac{1}{2}\mu_0 \frac{H_0}{M_0}(M_0\theta(t))^2 = \frac{1}{2}\mu_0 \frac{H_0}{M_0} m^2(t),$$

and

$$\langle w_z(t) \rangle = \frac{\mu_0}{4} \frac{H_0}{M_0} |\overline{m}|^2 .$$

Problem 7.5 The time-averaged magnetic energy density in a medium whose permeability is a function of frequency is given by[1]

$$
\begin{aligned}
\langle w_m(t) \rangle &= \frac{1}{4} \mathbf{h}^* \cdot \frac{\partial(\omega \overline{\mu})}{\partial \omega} \cdot \mathbf{h} \\
&= \frac{1}{4} \mathbf{b} \cdot \mathbf{h}^* + \frac{\omega}{4} \mathbf{h}^* \cdot \frac{\partial \overline{\mu}}{\partial \omega} \cdot \mathbf{h}.
\end{aligned}
\tag{7.22}
$$

Using the permeability tensor (4.13)

$$
\overline{\mu} = \mu_0 \begin{bmatrix} 1 + \chi & -i\kappa & 0 \\ i\kappa & 1 + \chi & 0 \\ 0 & 0 & 1 \end{bmatrix}
\tag{7.23}
$$

and the tensor elements (3.63a) and (3.63b)

$$
\chi = \frac{\omega_0 \omega_M}{\omega_0^2 - \omega^2} \quad \text{and} \quad \kappa = \frac{\omega \omega_M}{\omega_0^2 - \omega^2} ,
\tag{7.24}
$$

show that the last term in the energy expression (7.22) can be expressed as

$$
\frac{\omega}{4} \mathbf{h}^* \cdot \frac{\partial \overline{\mu}}{\partial \omega} \cdot \mathbf{h} = i \frac{\mu_0}{4} \frac{\omega}{\omega_M} (\mathbf{m} \times \mathbf{m}^*) \cdot \hat{\mathbf{z}}.
\tag{7.25}
$$

It can be shown that the volume integral of the first term in (7.22) vanishes [11,12]. Hence, the linear dependence on frequency in (7.25) suggests that we can write

$$
\langle w_m(t) \rangle = n_{\text{eff}} \hbar \omega,
\tag{7.26}
$$

where n_{eff} is an effective *quasi-particle number density* defined by

$$
n_{\text{eff}} = i \frac{\mu_0}{4 \hbar \omega_M} (\mathbf{m} \times \mathbf{m}^*) \cdot \hat{\mathbf{z}}.
\tag{7.27}
$$

---

[1]$\mathbf{b} \cdot \mathbf{h}^*$ is not the correct expression for the time-averaged magnetic energy density in a dispersive medium. Refer to the discussion by Fishman and Morganthaler [10] and by Stancil and Prabhakar [9, Sect. 7.4].

Solution 7.5

From (7.25), we must show that

$$\overline{h}^* \cdot \frac{\partial \overline{\overline{\mu}}}{\partial \omega} \cdot \overline{h} = i \frac{\mu_0}{\omega_M} \left(\overline{m} \times \overline{m}^* \right) \cdot \hat{z}. \tag{7.28}$$

First, we expand the derivative of the permeability using (7.23)

$$\frac{\partial \overline{\overline{\mu}}}{\partial \omega} = \mu_0 \frac{\partial}{\partial \omega} \begin{bmatrix} 1 + \chi & -i\kappa & 0 \\ i\kappa & 1 + \chi & 0 \\ 0 & 0 & 1 \end{bmatrix}$$

$$= \mu_0 \begin{bmatrix} \frac{\partial \chi}{\partial \omega} & -i \frac{\partial \kappa}{\partial \omega} & 0 \\ i \frac{\partial \kappa}{\partial \omega} & \frac{\partial \chi}{\partial \omega} & 0 \\ 0 & 0 & 0 \end{bmatrix}.$$

Since h_z is decoupled, we can work with the 2×2 tensor

$$\frac{\partial \overline{\overline{\mu}}}{\partial \omega} = \mu_0 \begin{bmatrix} \frac{\partial \chi}{\partial \omega} & -i \frac{\partial \kappa}{\partial \omega} \\ i \frac{\partial \kappa}{\partial \omega} & \frac{\partial \chi}{\partial \omega} \end{bmatrix}.$$

We now evaluate the tensor elements using (7.24)

$$\frac{\partial \chi}{\partial \omega} = \frac{\partial}{\partial \omega} \left[\frac{\omega_0 \omega_M}{\omega_0^2 - \omega^2} \right]$$

$$= \frac{2}{\omega_M} \left[\frac{\omega_0 \omega_M}{\omega_0^2 - \omega^2} \right] \left[\frac{\omega \omega_M}{\omega_0^2 - \omega^2} \right]$$

$$= \frac{2}{\omega_M} \chi \kappa.$$

Similarly

$$\frac{\partial \kappa}{\partial \omega} = \frac{\partial}{\partial \omega} \left[\frac{\omega \omega_M}{\omega_0^2 - \omega^2} \right]$$

$$= \frac{\omega_M}{\omega_0^2 - \omega^2} \left[1 + \frac{2\omega^2}{\omega_0^2 - \omega^2} \right]$$

$$= \frac{\omega_M}{\omega_0^2 - \omega^2} \left[\frac{\omega_0^2 - \omega^2 + 2\omega^2}{\omega_0^2 - \omega^2} \right]$$

$$= \frac{\omega_M}{\omega_0^2 - \omega^2} \left[\frac{\omega_0^2}{\omega_0^2 - \omega^2} + \frac{\omega^2}{\omega_0^2 - \omega^2} \right]$$

$$= \frac{1}{\omega_M} \left[\frac{\omega_0^2 \omega_M^2}{\left(\omega_0^2 - \omega^2\right)^2} + \frac{\omega^2 \omega_M^2}{\left(\omega_0^2 - \omega^2\right)^2} \right]$$

$$= \frac{1}{\omega_M} \left[\chi^2 + \kappa^2 \right].$$

With these results, the LHS of (7.28) becomes

$$\bar{h}^* \cdot \frac{\partial \bar{\bar{\mu}}}{\partial \omega} \cdot \bar{h} = \begin{bmatrix} h_x^* & h_y^* \end{bmatrix} \frac{\mu_0}{\omega_M} \begin{bmatrix} 2\chi\kappa & -i\left[\chi^2 + \kappa^2\right] \\ i\left[\chi^2 + \kappa^2\right] & 2\chi\kappa \end{bmatrix} \begin{bmatrix} h_x \\ h_y \end{bmatrix}$$

$$= \frac{\mu_0}{\omega_M} \left[2\chi\kappa h_x^* h_x - i\left(\chi^2 + \kappa^2\right) h_x^* h_y + i\left(\chi^2 + \kappa^2\right) h_x h_y^* + 2\chi\kappa h_y^* h_y \right]$$

$$= i\frac{\mu_0}{\omega_M} \left[\left(\chi h_x - i\kappa h_y\right)\left(-i\kappa h_x^* + \chi h_y^*\right) - \left(i\kappa h_x + \chi h_y\right)\left(\chi h_x^* + i\kappa h_y^*\right) \right]$$

$$(7.29)$$

Recall the relationship (3.62) between the small-signal magnetization and magnetic field intensity,

$$\bar{m} = \bar{\bar{\chi}} \cdot \bar{h} = \begin{bmatrix} \chi & -i\kappa \\ i\kappa & \chi \end{bmatrix} \begin{bmatrix} h_x \\ h_y \end{bmatrix}$$

or

$$\begin{bmatrix} m_x \\ m_y \end{bmatrix} = \begin{bmatrix} \chi h_x - i\kappa h_y \\ i\kappa h_x + \chi h_y \end{bmatrix}.$$

Comparing these expressions for m_x, m_y with the factored expression (7.29) gives

$$\bar{h}^* \cdot \frac{\partial \bar{\bar{\mu}}}{\partial \omega} \cdot \bar{h} = \frac{i\mu_0}{\omega_M} \left[m_x m_y^* - m_y m_x^* \right].$$

The RHS can be recognized as the z component of $\bar{m} \times \bar{m}^*$. We conclude that

$$\boxed{\frac{\omega}{4} \bar{h}^* \cdot \frac{\partial \bar{\bar{\mu}}}{\partial \omega} \cdot \bar{h} = i\frac{\mu_0}{4} \frac{\omega}{\omega_M} (\bar{m} \times \bar{m}^*) \cdot \hat{z}.}$$

Problem 7.6 Substitute the trial potential function

$$\psi = \begin{cases} \cos(\pi z/d), & |z| < d/2 \\ 0, & |z| > d/2 \end{cases} \tag{7.30}$$

into the variational formula

$$k^2 = \frac{\int \psi_z^* \psi_z dz}{\int [-(1+\chi)]\psi^* \psi dz} \tag{7.31}$$

and perform the necessary integrations to obtain the approximate dispersion relation

$$kd = \frac{\pi}{\sqrt{-(1+\chi)}}. \tag{7.32}$$

Solution 7.6

Consider the trial potential function

$$\psi = \begin{cases} \cos(\pi z/d), & |z| \leq d/2 \\ 0, & |z| > d/2 \end{cases}$$

The stationary expression for the dispersion relation is

$$k^2 = \frac{\int \psi_z^* \psi_z dz}{\int [-(1+\chi)]\psi^* \psi dz}.$$

Substituting the trial potential and carrying out the integrations gives

$$k^2 = \frac{\left(\frac{\pi}{d}\right)^2 \int_{-d/2}^{d/2} \sin^2\left(\frac{\pi z}{d}\right) dz}{-(1+\chi) \int_{-d/2}^{d/2} \cos^2\left(\frac{\pi z}{d}\right) dz}$$

$$= \frac{\frac{\pi}{d} \int_{-\pi/2}^{\pi/2} \sin^2 \theta d\theta}{-(1+\chi)\left(\frac{d}{\pi}\right) \int_{-\pi/2}^{\pi/2} \cos^2 \theta d\theta}.$$

Note that

$$\int_{-\pi/2}^{\pi/2} \sin^2 \theta d\theta = \int_{-\pi/2}^{\pi/2} \cos^2 \theta d\theta = \frac{\pi}{2},$$

so

$$k^2 = \left(\frac{\pi}{d}\right)^2 \frac{1}{-(1+\chi)},$$

or

$$\boxed{kd = \frac{\pi}{\sqrt{-(1+\chi)}}.}$$

Problem 7.7 In Chap. 5, we presented the exact forward volume wave dispersion relations with and without a ground plane ((5.50) and (5.43), respectively)

$$-\tan\left[kd\sqrt{-(1+\chi)}\right] = \frac{2\sqrt{-(1+\chi)}}{2+\chi(1+e^{-2kt})}, \tag{7.33}$$

$$\tan\left[\frac{k_t d}{2}\sqrt{-(1+\chi)}\right] = \frac{1}{\sqrt{-(1+\chi)}}, \tag{7.34}$$

respectively. Show that these relations reduce to (7.32) in the limit $[-(1+\chi)] \to 0$.

Solution 7.7

As was discussed in Problem 5.5, the forward volume wave dispersion relation with a ground plane is:

$$-\tan\left[kd\sqrt{-(1+\chi)}\right] = \frac{2\sqrt{-(1+\chi)}}{2+\chi(1+e^{-2kt})}.$$

Clearly, as $-(1+\chi) \to 0$ the RHS also goes to zero. The tangent on the LHS goes to zero when $kd\sqrt{-(1+\chi)} = n\pi$, or

$$kd = \frac{n\pi}{\sqrt{-(1+\chi)}}.$$

Since we know that $k \to \infty$ when $-(1+\chi) \to 0$, the lowest non-trivial solution is for $n = 1$. This gives the desired result

$$\boxed{kd = \frac{\pi}{\sqrt{-(1+\chi)}}.}$$

The dispersion relation without a ground plane is

$$\tan\left[\frac{kd}{2}\sqrt{-(1+\chi)}\right] = \frac{1}{\sqrt{-(1+\chi)}}.$$

In this case, the RHS increases without bound when $-(1+\chi) \to 0$. Consequently, we are interested in poles of the tangent function. These occur when

$$\frac{kd}{2}\sqrt{-(1+\chi)} = \frac{(2n-1)\pi}{2}, \quad n = 1, 2, 3 \ldots$$

or,

$$kd = \frac{(2n-1)\pi}{\sqrt{-(1+\chi)}}, \quad n = 1, 2, 3 \ldots.$$

The lowest-order positive solution is with $n = 1$, leading again to

$$\boxed{kd = \frac{\pi}{\sqrt{-(1+\chi)}}.}$$

References

1. W.F. Brown, Jr., *Micromagnetics*, Series Interscience Tracts on Physics and Astronomy, vol. 18 (Interscience Publishers, New York, 1963)
2. D.D. Stancil, Variational formulation of magnetostatic wave dispersion relations. IEEE Trans. Magn. **19**, 1865 (1983)
3. N.E. Buris, D.D. Stancil, Magnetostatic surface-wave propagation in ferrite thin films with arbitrary variations of the magnetization through the film thickness, in *Transactions on Microwave Theory and Techniques*, vol. MTT-33 (1985), p. 484
4. N.E. Buris, D.D. Stancil, Magnetostatic volume modes of ferrite thin films with magnetization inhomogeneities through the film thickness, in *IEEE Transactions on Microwave Theory and Techniques*, vol. MTT-33 (1985), p. 1089
5. N.E. Buris, D.D. Stancil, Magnetostatic backward waves in low dose ion implanted YIG films. IEEE Trans. Magn. **22**, 859 (1986)
6. M. Tsutsumi, Y. Masaoka, T. Ohira, N. Kumagai, A new technique for magnetostatic wave delay lines. IEEE Trans. Microwave Theory Tech. **29**, 583 (1981)
7. E. Sawado, N.S. Chang, Variational approach to analysis of propagation of magnetostatic waves in highly inhomogeneously magnetized media. J. Appl. Phys. **55**, 1062 (1984)
8. J. Matthews, R.L. Walker, *Mathematical Methods of Physics* (W. A. Benjamin Inc., Menlo Park, 1970)
9. D.D. Stancil, A. Prabhakar, *Spin Waves: Theory and Applications* (Springer, New York, 2009)
10. D.A. Fishman, F.R. Morgenthaler, Investigation of the velocity of energy circulation of magnetostatic modes in ferrites. J. Appl. Phys **54**, 3387 (1983)
11. F.R. Morgenthaler, Dynamic magnetoelastic coupling in ferromagnets and antiferromagnets. IEEE Trans. Magn. **8**, 130 (1972)
12. F.R. Morgenthaler, Control of magnetostatic waves in thin films by means of spatially nonuniform bias fields. Circuits Systems Signal Process **4**, 63 (1985)

Optical-Spin Wave Interactions

8

In this chapter, we consider thin films that act simultaneously as waveguides for optical modes as well as spin wave modes. We first treat the optical modes of a thin-film waveguide consisting of a nonmagnetic isotropic dielectric. These modes are explored in Problems 8.1–8.6. We next discuss the nature of the magneto-optic interaction and how it affects the constitutive tensors. In particular, Problems 8.7–8.9 consider the relationship between modeling magneto-optic phenomena using either the permittivity or permeability tensors, and Problem 8.10 considers the phenomenon of Faraday rotation using a gyrotropic permittivity tensor. Finally, Problems 8.11–8.15 explore coupled-mode theory as applied to the interaction between optical guided modes and spin waves owing to the perturbation to the optical tensors caused by the presence of dipolar spin waves.

8.1 Optical Guided Modes of a Dielectric Thin Film

The normal modes of electromagnetic wave propagation in a waveguide are orthogonal, i.e., they do not couple to one another. However, perturbations in the medium can introduce coupling between otherwise uncoupled modes.[1] In the presence of such coupling, energy that starts off in one mode will "gradually" convert to the second mode, and then back to the original mode as the modes propagate. In this context, "gradually" means that the scale over which the mode conversion occurs is large compared to the wavelengths of the propagating modes. The perturbations that we are particularly interested in are those arising from the co-propagation of spin waves in the film.

---

[1]Treatments of the optical guided modes of dielectric thin films can be found in Yariv and Yeh [1], Yariv [2], and Haus [3].

Fig. 8.1 Geometry of a symmetric dielectric waveguide. (Reproduced from [4] with permission.)

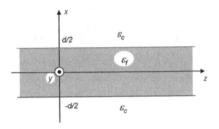

We have spent some time understanding the spin wave modes of a magnetic thin film, so now we turn our attention to the *optical modes*. We will then be in a position to consider how the mode conversion occurs. We begin by considering the optical modes of the symmetric geometry shown in Fig. 8.1.

The general approach to finding the *guided modes* is to create a set of trial solution functions that represent propagating waves inside the film, and exponentially decaying fields outside of the film. The amplitudes of the functions in each region are then found to ensure that the boundary conditions of tangential **E** and **H** continuous are satisfied at the film boundaries. The solutions can be divided into two classes: those with the electric field perpendicular to the direction of propagation (*Transverse Electric*, or TE), and modes with the magnetic field perpendicular to the direction of propagation (*Transverse Magnetic*, or TM).

8.1.1 Transverse Electric Modes

For definiteness, we will consider propagation along the $\hat{\mathbf{z}}$ direction. The transverse electric field is then in the $\hat{\mathbf{y}}$ direction, and the magnetic field has components along x and z. We are particularly interested in the y and z components of the fields, since these quantities must be chosen to satisfy the boundary conditions at $z = \pm d/2$.

The fields for the even TE modes are given by

$$E_y = e_y(x)e^{ik_z z}, \tag{8.1}$$

$$H_z = h_z(x)e^{ik_z z}, \tag{8.2}$$

where

$$e_y(x) = \begin{cases} Ae^{-\alpha_x(x-d/2)}\cos(k_x d/2), & x > d/2, \\ A\cos(k_x x), & |x| \leq d/2, \\ Ae^{\alpha_x(x+d/2)}\cos(k_x d/2), & x < -d/2, \end{cases} \tag{8.3}$$

and

$$
h_z(x) = \begin{cases}
\dfrac{i\alpha_x}{\omega\mu_0} A e^{-\alpha_x(x-d/2)} \cos(k_x d/2), & x > d/2, \\[2ex]
\dfrac{i A k_x}{\omega\mu_0} \sin(k_x x), & |x| \le d/2, \\[2ex]
-\dfrac{i\alpha_x}{\omega\mu_0} A e^{\alpha_x(x+d/2)} \cos(k_x d/2), & x < -d/2.
\end{cases}
\tag{8.4}
$$

The parameters α_x and k_x satisfy the dispersion relations for plane waves in the two media:

$$
\omega^2 \mu_0 \varepsilon_c = k_z^2 - \alpha_x^2,
\tag{8.5a}
$$
$$
\omega^2 \mu_0 \varepsilon_f = k_z^2 + k_x^2.
\tag{8.5b}
$$

Eliminating k_z from (8.5) gives the relation between k_x and α_x:

$$
k_x^2 = \omega^2 \mu_0 (\varepsilon_f - \varepsilon_c) - \alpha_x^2.
\tag{8.6}
$$

Imposing the boundary conditions at $x = \pm d/2$ on h_z and e_y leads to the following transcendental equation for k_x given a specific value of ω:

$$
\tan\left[\frac{k_x d}{2}\right] = \sqrt{\frac{\omega^2 \mu_0 (\varepsilon_f - \varepsilon_c)}{k_x^2} - 1}.
\tag{8.7}
$$

The wave number k_z for propagation along the $\hat{z}$ direction can then be obtained from (8.5b).

Starting from a similar field structure but with $e_y(x) = A \sin(k_x x)$ inside the film, the transcendental equation for the antisymmetric TE modes is found to be (cf. Problem 8.3)

$$
\cot\left[\frac{k_x d}{2}\right] = -\sqrt{\frac{\omega^2 \mu_0 (\varepsilon_f - \varepsilon_c)}{k_x^2} - 1}.
\tag{8.8}
$$

For both symmetric and antisymmetric modes, the modes only propagate as guided modes above a specific cutoff frequency given by

$$
f_{cm} = \frac{m}{2d\sqrt{\mu_0(\varepsilon_f - \varepsilon_c)}},
\tag{8.9}
$$

where m is even for symmetric modes, and odd for antisymmetric modes.

8.1.2 Transverse Magnetic Modes

The fields for symmetric TM modes are given by

$$H_y = h_y(x)e^{ik_z z}, \tag{8.10}$$

$$E_z = e_z(x)e^{ik_z z}, \tag{8.11}$$

where

$$h_y(x) = \begin{cases} Ae^{-\alpha_x(x-d/2)}\cos{(k_x d/2)}, & x > d/2, \\ A\cos(k_x x), & |x| \le d/2, \\ Ae^{\alpha_x(x+d/2)}\cos{(k_x d/2)}, & x < -d/2, \end{cases} \tag{8.12}$$

$$e_z(x) = \begin{cases} -\dfrac{i\alpha_x}{\omega\varepsilon_c}Ae^{-\alpha_x(x-d/2)}\cos{(k_x d/2)}, & x > d/2, \\ -\dfrac{iAk_x}{\omega\varepsilon_f}\sin(k_x x), & |x| \le d/2, \\ \dfrac{i\alpha_x}{\omega\varepsilon_c}Ae^{\alpha_x(x+d/2)}\cos{(k_x d/2)}, & x < -d/2. \end{cases} \tag{8.13}$$

After matching boundary conditions, the dispersion relation for the symmetric modes is given by

$$\tan\left[\frac{k_x d}{2}\right] = \frac{\varepsilon_f}{\varepsilon_c}\sqrt{\frac{\omega^2\mu_0(\varepsilon_f - \varepsilon_c)}{k_x^2} - 1}. \tag{8.14}$$

Again, the modes with antisymmetric h_y can be found by starting with $h_y(x) = A\sin{(k_x x)}$ inside the film and matching boundary conditions. The antisymmetric mode dispersion relation is found to be

$$\cot\left[\frac{k_x d}{2}\right] = -\frac{\varepsilon_f}{\varepsilon_c}\sqrt{\frac{\omega^2\mu_0(\varepsilon_f - \varepsilon_c)}{k_x^2} - 1}. \tag{8.15}$$

As with the TE modes, one first solves (8.14) or (8.15) for k_x, then the propagating wave number k_z is obtained from (8.5b). The *cutoff frequencies* for the TM modes are the same as for the TE modes.

Problem 8.1 Near cutoff, the propagation characteristics of guided modes in a symmetric dielectric waveguide are dominated by the outside medium as indicated by

$$k_z^2 \approx \omega\mu_0\varepsilon_c. \tag{8.16}$$

Starting with the requirement that $\alpha_x \to 0$, show that (8.16) is valid for both even and odd TE and TM modes near cutoff.

Solution 8.1

From (8.5b), we have

$$k_z^2 = \omega^2 \mu_0 \varepsilon_f - k_x^2,$$

while from (8.6), we have

$$k_x^2 = \omega^2 \mu_0 (\varepsilon_f - \varepsilon_c) - \alpha_x^2.$$

Eliminating k_x from the above two equations gives

$$k_z^2 = \omega^2 \mu_0 \varepsilon_f - \omega^2 \mu_0 (\varepsilon_f - \varepsilon_c) + \alpha_x^2$$
$$= \omega^2 \mu_0 \varepsilon_c + \alpha_x^2.$$

As $\alpha_x \to 0$, we get

$$\boxed{k_z^2 \approx \omega^2 \mu_0 \varepsilon_c} \qquad (8.17)$$

so that the propagation is dominated by the outside medium (ε_c).

Note that these conclusions are valid for all modes, i.e., both even and odd TE and TM. The specific dispersion relation for each mode determines the precise value of k_x for a given ω, but the basic dispersion relations for plane waves in each medium require that

$$\lim_{\alpha_x \to 0} k_x^2 = \omega^2 \mu_0 (\varepsilon_f - \varepsilon_c)$$

and (8.17) follows directly from this result.

Problem 8.2 Each branch of the tangent function in the dispersion relation for symmetric TE modes (8.7)

$$\tan\left[\frac{k_x d}{2}\right] = \sqrt{\frac{\omega^2 \mu_0 (\varepsilon_f - \varepsilon_c)}{k_x^2} - 1} \qquad (8.18)$$

yields a separate mode solution. On each positive branch, the tangent ranges from 0 to ∞ while $k_x d/2$ is bounded between $n\pi$ and $(n + 1/2)\pi$. If k_x is bounded, the right-hand side of (8.18) can be singular only if $\omega \to \infty$. Take the limit $\omega \to \infty$ with k_x finite in (8.5b)

$$\omega^2 \mu_0 \varepsilon_f = k_z^2 + k_x^2 \qquad (8.19)$$

to obtain the dispersion relation far from cutoff

$$k_z^2 \approx \omega^2 \mu_0 \varepsilon_f. \qquad (8.20)$$

Is this conclusion limited to symmetric TE modes, or does it apply to both TE and TM modes regardless of symmetry? Explain your answer.

Solution 8.2

From (8.19), we have

$$\omega^2 \mu_0 \varepsilon_f = k_z^2 + k_x^2.$$

If k_x is bounded, but $\omega \to \infty$, then k_z must become arbitrarily large also. In this case

$$\boxed{\omega^2 \mu_0 \varepsilon_f \simeq k_z^2.} \tag{8.21}$$

We conclude that far from cut off, the mode is dominated by the property of the film layer.

Equation (8.19) is valid for any plane wave in the film. Consequently, the dispersion relation (8.21) is valid for both TE and TM guided modes.

Problem 8.3 Derive the antisymmetric TE mode dispersion relation (8.8)

$$\cot\left[\frac{k_x d}{2}\right] = -\sqrt{\frac{\omega^2 \mu_0 (\varepsilon_f - \varepsilon_c)}{k_x^2} - 1} \tag{8.22}$$

starting with the TE mode function

$$e_y(x) = \begin{cases} Ce^{-\alpha_x x}, & x > d/2, \\ A\sin(k_x x), & |x| \le d/2, \\ -Ce^{+\alpha_x x}, & x < -d/2. \end{cases} \tag{8.23}$$

Solution 8.3

Following the geometry in Fig. 8.1, consider the electric field of the form given to us

$$e_y(x) = \begin{cases} Ce^{-\alpha_x x} & x > d/2, \\ A\sin k_x x & |x| \le d/2, \\ -Ce^{\alpha_x x} & x < -d/2. \end{cases} \tag{8.23}$$

We can use Maxwell's equation from Faraday's law to find $h_z(x)$

$$\nabla \times \overline{e} = i\omega \mu_0 \overline{h} \implies h_z = -\frac{i}{\omega \mu_0}\frac{\partial e_y}{\partial x}$$

Therefore

$$h_z = \begin{cases} +\dfrac{i\alpha_x}{\omega \mu_0} Ce^{-\alpha_x x} & x > d/2, \\[2mm] -\dfrac{ik_x}{\omega \mu_0} A\cos k_x x & |x| \le d/2, \\[2mm] \dfrac{i\alpha_x}{\omega \mu_0} Ce^{\alpha_x x} & x < -d/2. \end{cases}$$

Now consider the boundary conditions that e_y and h_z are continuous (symmetry allows us to consider only $x = d/2$), to get

$$Ce^{-\alpha_x d/2} = A \sin \frac{k_x d}{2}, \qquad (8.24a)$$

$$\frac{i\alpha_x}{\omega\mu_0} Ce^{-\alpha_x d/2} = -\frac{ik_x}{\omega\mu_0} A \cos \frac{k_x d}{2}. \qquad (8.24b)$$

Equations (8.24a) and (8.24b) are easily solved to find

$$\alpha_x \sin \frac{k_x d}{2} = -k_x \cos \frac{k_x d}{2}$$

$$\text{or,} \qquad \cot \frac{k_x d}{2} = -\frac{\alpha_x}{k_x}.$$

From (8.6)

$$k_x^2 = \omega^2 \mu_0 (\varepsilon_f - \varepsilon_c) - \alpha_x^2,$$

from which we conclude that

$$\frac{\alpha_x}{k_x} = \sqrt{\frac{\omega^2 \mu_0 (\varepsilon_f - \varepsilon_e)}{k_x^2} - 1}.$$

Hence, we get

$$\boxed{\cot \left[\frac{k_x d}{2}\right] = -\sqrt{\frac{\omega^2 \mu_0 (\varepsilon_f - \varepsilon_e)}{k_x^2} - 1.}}$$

Problem 8.4 Derive the antisymmetric TM mode dispersion relation (8.15)

$$\cot \left[\frac{k_x d}{2}\right] = -\frac{\varepsilon_f}{\varepsilon_c} \sqrt{\frac{\omega^2 \mu_0 (\varepsilon_f - \varepsilon_c)}{k_x^2} - 1} \qquad (8.25)$$

starting with a TM mode function h_y with the same functional form as e_y in (8.23).

Solution 8.4

For antisymmetric modes, the magnetic field intensity can be written as

$$H_y = h_y(x) \exp(ik_z z)$$

where

$$h_y(x) = \begin{cases} Ce^{-\alpha_x x}, & x > d/2, \\ A \sin(k_x x), & |x| \le d/2, \\ -Ce^{\alpha x}, & x < -d/2. \end{cases}$$

The electric field can be obtained from Ampere's law

$$E_z = \frac{i}{\omega\varepsilon}\frac{\partial H_y}{\partial x}$$

where the electric field is of the form

$$E_z = e_z(x)\exp(ik_z z)$$

and

$$e_z(x) = \begin{cases} \dfrac{-iC\alpha_x}{\omega\varepsilon_c}e^{-\alpha_x x}, & x > d/2, \\[2mm] \dfrac{iAk_x}{\omega\varepsilon_f}\cos(k_x x), & |x| \le d/2, \\[2mm] \dfrac{-iC\alpha_x}{\omega\varepsilon_c}e^{\alpha x}, & x < -d/2. \end{cases}$$

Matching the boundary conditions at $x = d/2$ gives

$$C\exp\left(-\frac{\alpha_x d}{2}\right) = A\sin\left(\frac{k_x d}{2}\right)$$

$$\frac{-iC\alpha_x\exp(-\alpha_x d/2)}{\omega\varepsilon_0} = \frac{iAk_x}{\omega\varepsilon_f}\cos\left(\frac{k_x d}{2}\right)$$

Combining these equations gives

$$\cot\left(\frac{k_x d}{2}\right) = -\frac{\alpha_x}{k_x}\frac{\varepsilon_f}{\varepsilon_c}.$$

Using (8.6), $\alpha_x^2 = \omega^2\mu_0(\varepsilon_f - \varepsilon_c) - k_x^2$, we obtain the desired result

$$\boxed{\cot\left[\frac{k_x d}{2}\right] = -\frac{\varepsilon_f}{\varepsilon_0}\sqrt{\frac{\omega^2\mu_0(\varepsilon_f - \varepsilon_0)}{k_x^2}}.}$$

Problem 8.5 Most thin-film waveguides encountered in optical devices are not symmetric about the guiding layer. The general three-layer geometry consists of a cover, a film, and a substrate as shown in Fig. 8.2.

The mode fields in this structure will be neither symmetric nor antisymmetric. Consequently, we choose a trial TE mode function of the form

$$e_y(x) = \begin{cases} Ae^{-\alpha_c x}, & x > 0, \\ B\cos(k_x x) + C\sin(k_x x), & -d \le x \le 0, \\ De^{+\alpha_s x}, & x < -d. \end{cases} \qquad (8.26)$$

Fig. 8.2 General three-layer dielectric waveguide consisting of a cover, a film, and a substrate. (Reproduced from [4] with permission.)

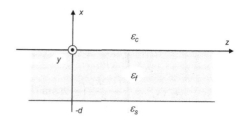

(a) Find the mode function for the z component of the magnetic field intensity h_z.

(b) By requiring that e_y and h_z be continuous at $x = 0, -d$, show that the TE mode dispersion relation is

$$\tan(k_x d) = \frac{k_x(\alpha_c + \alpha_s)}{k_x^2 - \alpha_c \alpha_s},\tag{8.27}$$

where

$$\alpha_c = \sqrt{k_z^2 - \omega^2 \mu_0 \varepsilon_c},\tag{8.28a}$$

$$\alpha_s = \sqrt{k_z^2 - \omega^2 \mu_0 \varepsilon_s},\tag{8.28b}$$

and k_x is defined as (8.5b)

$$\omega^2 \mu_0 \varepsilon_f = k_z^2 + k_x^2.\tag{8.28c}$$

Solution 8.5

(a) We choose a trial expression for the electric field of the form

$$E_y = e_y(x)\exp(ik_z z)$$

where from (8.26)

$$e_y(x) = \begin{cases} Ae^{-\alpha_c x}, & x > 0, \\ B\cos(k_x x) + C\sin(k_x x), & -d \le x \le 0, \\ De^{\alpha_s x}, & x < -d. \end{cases}\tag{8.29}$$

The magnetic field intensity can be obtained from Faraday's law

$$H_z = \frac{-i}{\omega \mu_0}\frac{\partial E_y}{\partial x}.$$

Taking the x derivative of (8.29) leads to

$$
h_z(x) = \begin{cases}
\dfrac{i\alpha_c}{\omega\mu_0} A e^{-\alpha_c x}, & x > 0 \\[2ex]
\dfrac{i - k_x}{\omega\mu_0}\Big(-B\sin(k_x x) + C\cos(k_x x)\Big), & -d \le x \le 0 \\[2ex]
\dfrac{-i\alpha_s}{\omega\mu_0} D e^{x}\alpha_s, & x < -d.
\end{cases}
$$

(b) Matching the boundary conditions that e_y and h_z are continuous at $x = 0$ give

$$A = B,$$

$$\frac{i\alpha_c}{\omega\mu_0} A = \frac{-ik_x}{\omega\mu_0} C.$$

It follows that

$$\Rightarrow A = B = \frac{-k_x}{\alpha_c} C.$$

The boundary conditions at $x = -d$ give

$$B\cos(k_x d) - C\sin(k_x d) = D\exp(-\alpha_s d) \tag{8.30a}$$

$$\frac{-k_x i}{\omega\mu_0}(B\sin(k_x d) + C\cos(k_x d)) = \frac{-iD}{\omega\mu_0}\exp(-\alpha_s d)\,\alpha_s \tag{8.30b}$$

Dividing (8.30a) by (8.30b), we get

$$\frac{B\cos(k_x d) - C\sin(k_x d)}{k_x(B\sin(k_x d) + C\cos(k_x d))} = \frac{1}{\alpha_s}$$

or,

$$k_x\left(\frac{-k_x}{\alpha_c}C\sin(k_x d) + C\cos(k_x d)\right) = -C\sin(k_x d)\,\alpha_s - \frac{k_x\alpha_s}{\alpha_c}C\cos(k_x d)$$

$$-k_x^2\sin(k_x d) + k_x\alpha_c\cos(k_x d) = \alpha_s\alpha_c\sin(k_x d) - k_x\alpha_s\cos(k_x d)$$

$$\therefore \quad \boxed{\tan(k_x d) = \frac{k_x(\alpha_c + \alpha_s)}{k_x^2 - \alpha_c\alpha_s}}$$

where

$$\alpha_c = \sqrt{k_z^2 - \omega^2\mu_0\varepsilon_c},$$

$$\alpha_s = \sqrt{k_z^2 - \omega^2\mu_0\varepsilon_s}.$$

Problem 8.6 Consider the TM modes of the general dielectric waveguide shown in Fig. 8.2. A suitable trial mode function is

$$
h_y(x) = \begin{cases} Ae^{-\alpha_c x}, & x > 0, \\ B\cos(k_x x) + C\sin(k_x x), & -d \leq x \leq 0, \\ De^{+\alpha_s x}, & x < -d. \end{cases} \tag{8.31}
$$

(a) Find the mode function for the z component of the electric field intensity e_z.
(b) By requiring that h_y and e_z be continuous at $x = 0, -d$, show that the TM mode dispersion relation is

$$
\tan(k_x d) = \frac{k_x(\tilde{\alpha}_c + \tilde{\alpha}_s)}{k_x^2 - \tilde{\alpha}_c \tilde{\alpha}_s}, \tag{8.32}
$$

where

$$
\tilde{\alpha}_c = \frac{\varepsilon_f}{\varepsilon_c}\alpha_c, \tag{8.33a}
$$

$$
\tilde{\alpha}_s = \frac{\varepsilon_f}{\varepsilon_s}\alpha_s, \tag{8.33b}
$$

and α_c, α_s, and k_x are defined as in Problem 8.5.

Solution 8.6

(a) We choose a trial solution for the magnetic field intensity of the form

$$
H_y = h_y(x)\exp(ik_z z)
$$

where from (8.31)

$$
h_y(x) = \begin{cases} Ae^{-\alpha_c x}, & x > 0, \\ C\sin(k_x x) + B\cos(k_x x), & d \leq x \leq 0, \\ De^{+\alpha_s x}, & x < -d. \end{cases}
$$

Assuming the electric field is of the form

$$
E_z = e_z(x)\exp(ik_z z),
$$

the electric field is obtained from the above magnetic field using

$$
E_z = \frac{i}{\omega\varepsilon}\frac{\partial H_y}{\partial x}.
$$

The result is

$$
e_z(x) = \begin{cases} \dfrac{-i\alpha_c}{\omega\varepsilon_c} A e^{-\alpha_c x}, & x > 0, \\[2ex] \dfrac{ik_x}{\omega\varepsilon_f}\left(-B\sin(k_x x) + C\cos(k_x x)\right), & -d \le x \le 0, \\[2ex] \dfrac{i\alpha_s}{\omega\varepsilon_s} D e^x \alpha_s, & x < -d. \end{cases}
$$

(b) Boundary conditions require that h_y and e_z must be continuous at $x = 0$ and $x = -d$. Applying the conditions at $x = 0$ gives

$$A = B$$

$$\frac{-i\alpha_c}{\omega\varepsilon_c} A = \frac{ik_x}{\omega\varepsilon_f} C$$

Or equivalently,

$$A = B = \frac{-k_x}{\alpha_c}\frac{\varepsilon_c}{\varepsilon_f} C \equiv -\frac{k_x}{\tilde{\alpha}_c} C.$$

Similarly, applying the boundary conditions at $x = -d$ gives

$$B\cos(k_x d) - C\sin(k_x d) = D e^{-\alpha_s d} \tag{8.34a}$$

$$\frac{k_x i}{\omega\varepsilon_f}\left(B\sin(k_x d) + C\cos(k_x d)\right) = \frac{iD}{\omega\varepsilon_s} e^{-\alpha_s d}\alpha_s \tag{8.34b}$$

Dividing (8.34a) by (8.34b), we get

$$\frac{B\cos(k_x d) - C\sin(k_x d)}{k_x\left(B\sin(k_x d) + C\cos(k_x d)\right)} = \frac{1}{\alpha_s}\frac{\varepsilon_s}{\varepsilon_f} \equiv \frac{1}{\tilde{\alpha}_s}$$

Therefore,

$$k_x\left(\frac{-k_x}{\tilde{\alpha}_c} C\sin(k_x d) + C\cos(k_x d)\right) = -C\tilde{\alpha}_s\sin(k_x d) - \frac{k_x\tilde{\alpha}_s}{\tilde{\alpha}_c} C\cos(k_x d)$$

$$\frac{-k_x^2}{\tilde{\alpha}_c}\sin(k_x d) + k_x\cos(k_x d) = -\tilde{\alpha}_s\sin(k_x d) - k_x\frac{\tilde{\alpha}_s}{\tilde{\alpha}_c}\cos(k_x d)$$

Thus

$$\boxed{\tan(k_x d) = \frac{k_x(\tilde{\alpha}_c + \tilde{\alpha}_s)}{k_x^2 - \tilde{\alpha}_c\tilde{\alpha}_s}} \tag{8.32}$$

where

$$\tilde{\alpha}_c = \frac{\varepsilon_f}{\varepsilon_c}\alpha_c, \tag{8.33a}$$

$$\tilde{\alpha}_s = \frac{\varepsilon_f}{\varepsilon_s}\alpha_s. \tag{8.33b}$$

8.2 Magnetic and Electric Susceptibilities

We now turn our attention from the optical modes in a dielectric thin film to material properties such as the magnetization $\mathbf{M}$ and electric polarization $\mathbf{P}$. (The connection between these two topics will be seen in the final set of problems pertaining to *coupled-mode theory*.) At "low" frequencies $\mathbf{M}$ is interpreted as the magnetic moment per unit volume. However, electric and magnetic responses are more difficult to separate at "high" frequencies. To see why this is, as well as what we mean by "low" and "high," consider the classical expression for the magnetic moment of a finite volume of magnetic material

$$\mu = \frac{1}{2} \int_V \mathbf{r} \times \mathbf{J} \, dv, \tag{8.35}$$

where $\mathbf{J}$ is given by

$$\mathbf{J} = \frac{\partial \mathbf{P}}{\partial t} + \nabla \times \mathbf{M}. \tag{8.36}$$

The integral in (8.35) is over a volume V that is larger than the volume of the sample V_0. Substituting (8.36) into (8.35) gives

$$\mu = \frac{1}{2} \int_V \mathbf{r} \times \frac{\partial \mathbf{P}}{\partial t} dv + \frac{1}{2} \int_V \mathbf{r} \times (\nabla \times \mathbf{M}) \, dv. \tag{8.37}$$

We can integrate the last term on the right by parts:

$$\int_V \mathbf{r} \times (\nabla \times \mathbf{M}) \, dv = - \oint_S \mathbf{r} \times (\mathbf{M} \times \mathbf{ds}) - \int_V (\mathbf{M} \times \nabla) \times \mathbf{r} \, dv. \tag{8.38}$$

Here the surface S encloses the volume V. Since V is larger than the sample, the surface is outside the sample so the magnetization $\mathbf{M}$ is zero. Consequently, the surface term vanishes. We can simplify the second term on the right-hand-side by using the identity (see Problem 8.8):

$$(\mathbf{M} \times \nabla) \times \mathbf{r} = -2\mathbf{M}. \tag{8.39}$$

Consequently, the expression for the magnetic moment (8.37) becomes

$$\mu = \frac{1}{2} \int_V \mathbf{r} \times \frac{\partial \mathbf{P}}{\partial t} dv + \int_V \mathbf{M} dv. \tag{8.40}$$

If $\mathbf{M}$ were rigorously the magnetic moment per unit volume, then only the second term would be present. Clearly the interpretation is not so straightforward if we cannot neglect the first term. We will see in Problem 8.9 that this term can only be neglected if the wavelength is large compared to the same volume V_0. While this is usually satisfied at microwave frequencies, it is not at optical frequencies. As a result, electric and optical properties are not easily separated at optical frequencies. We explore these ideas further in Problems 8.7–8.10.

Problem 8.7 In this problem, we will show that an infinite medium that is both electrically and magnetically anisotropic can be alternatively modeled with a single effective permittivity tensor for uniform plane waves.

(a) Starting with the constitutive relations

$$M(\omega) = \overline{\overline{\chi}}_m(\omega) \cdot H(\omega), \tag{8.41}$$

$$B = \overline{\overline{\mu}} \cdot H, \tag{8.42}$$

$$\overline{\overline{\mu}} = \mu_0(\overline{\overline{I}} + \overline{\overline{\chi}}_m), \tag{8.43}$$

show that the relation between M and B is

$$M = \frac{1}{\mu_0}\overline{\overline{\chi}}'_m \cdot B, \tag{8.44}$$

where

$$\overline{\overline{\chi}}'_m = \overline{\overline{\chi}}_m \cdot \left[\overline{\overline{I}} + \overline{\overline{\chi}}_m\right]^{-1}. \tag{8.45}$$

(b) From the discussions in Chapter 4, and the treatment in [4, Sect. 4.6], we know that for uniform plane waves, Maxwell's equations

$$\nabla \times E = -\frac{\partial B}{\partial t}, \tag{8.46}$$

$$\nabla \times B = \frac{1}{c^2}\frac{\partial E}{\partial t} + \mu_0 J, \tag{8.47}$$

can be written

$$\overline{\overline{k}} \cdot E = \omega B, \tag{8.48}$$

$$\overline{\overline{k}} \cdot B = -\frac{\omega}{c^2}E - i\mu_0 J, \tag{8.49}$$

where

$$J = -i\omega P + i\overline{\overline{k}} \cdot M, \tag{8.50}$$

and $\overline{\overline{k}}$ is defined by

$$\overline{\overline{k}} \equiv k \times \overline{\overline{I}} = \begin{bmatrix} 0 & -k_z & k_y \\ k_z & 0 & -k_x \\ -k_y & k_x & 0 \end{bmatrix}. \tag{8.51}$$

Use (8.44), (8.48), and (8.49) to eliminate M and B from the expression for the current (8.50). Finally, use the constitutive relation

$$P(\omega) = \varepsilon_0 \overline{\overline{\chi}}_e(\omega) \cdot E(\omega) \tag{8.52}$$

to express $\mathbf{J}$ entirely in terms of $\mathbf{E}$. Show that $\mathbf{J}$ can be obtained from an effective susceptibility tensor defined as

$$\mathbf{J} = -i\omega\varepsilon_0 \, \overline{\overline{\chi}}_e^{\text{eff}} \cdot \mathbf{E} \tag{8.53}$$

where

$$\overline{\overline{\chi}}_e^{\text{eff}} = \overline{\overline{\chi}}_e - \frac{c^2}{\omega^2} \mathbf{k} \cdot \overline{\overline{\chi}}'_m \cdot \mathbf{k} \,. \tag{8.54}$$

(c) Show that (8.53) reduces to

$$\mathbf{J} = -i\omega\varepsilon_0 \frac{(\chi_e + \chi'_m)}{(1 - \chi'_m)} \mathbf{E} \equiv -i\omega\varepsilon_0 \chi_e^{(\text{eff})} \mathbf{E} \tag{8.55}$$

for an electrically and magnetically isotropic medium. (Hint: Use (8.49) along with (8.48) to show that $\overline{\mathbf{k}} \cdot \overline{\mathbf{k}} \cdot \mathbf{E} = -\frac{\omega^2}{c^2}\mathbf{E} - i\omega\mu_0\mathbf{J}$.)

Solution 8.7

(a) We begin with the constitutive relations

$$\overline{M}(\omega) = \overline{\overline{\chi}}_m(\omega) \cdot \overline{H}(\omega)$$
$$\overline{B} = \overline{\overline{\mu}} \cdot \overline{H}$$
$$\overline{\overline{\mu}} = \mu_0 \left(\overline{\overline{I}} + \overline{\overline{\chi}}_m\right)$$
$$\overline{B} = \mu_0 \left(\overline{\overline{I}} + \overline{\overline{\chi}}_m\right) \cdot \overline{H}$$
$$\overline{H} = \frac{1}{\mu_0} \left(\overline{\overline{I}} + \overline{\overline{\chi}}_m\right)^{-1} \cdot \overline{B}$$

or

$$\overline{\overline{\chi}}_m \cdot \overline{H} = \frac{1}{\mu_0} \overline{\overline{\chi}}_m \cdot \left(\overline{\overline{I}} + \overline{\overline{\chi}}_m\right)^{-1} \cdot \overline{B} = \frac{1}{\mu_0} \overline{\overline{\chi}}'_m \cdot \overline{B}.$$

We conclude that

$$\boxed{\overline{M} = \frac{1}{\mu_0} \overline{\overline{\chi}}'_m \cdot \overline{B}, \qquad \overline{\overline{\chi}}'_m = \overline{\overline{\chi}}_m \cdot \left(\overline{\overline{I}} + \overline{\overline{\chi}}_m\right)^{-1}.}$$

(b) We begin with the constitutive relations

$$\overline{M} = \frac{1}{\mu_0} \overline{\overline{\chi}}'_m \cdot \overline{B} \tag{8.56}$$
$$\overline{P} = \varepsilon_0 \overline{\overline{\chi}}_e \cdot \overline{E} \tag{8.57}$$

and Maxwell's equation from Faraday's law for plane waves

$$\overline{\overline{k}} \cdot \overline{E} = \omega \overline{B}. \tag{8.58}$$

We proceed to eliminate $\overline{M}$ and $\overline{B}$ from the current density expression

$$\overline{J} = -i\omega \overline{P} + i\overline{\overline{k}} \cdot \overline{M}. \tag{8.59}$$

Using (8.56) in the last term of (8.59) gives

$$\overline{\overline{k}} \cdot \overline{M} = \overline{\overline{k}} \cdot \overline{\overline{\chi}}'_{\mathrm{m}} \cdot \frac{\overline{B}}{\mu_0}.$$

Using (8.58) to eliminate $\overline{B}$ gives

$$\overline{\overline{k}} \cdot \overline{M} = \overline{\overline{k}} \cdot \overline{\overline{\chi}}'_{\mathrm{m}} \cdot \overline{\overline{k}} \cdot \overline{E} \frac{1}{\omega \mu_0}.$$

Using this expression along with (8.57) in (8.59) gives

$$\begin{aligned}
\overline{J} &= -i\omega\varepsilon_0 \overline{\overline{\chi}}_{\mathrm{e}} \cdot \overline{E} + \frac{i}{\omega\mu_0}\overline{\overline{k}} \cdot \overline{\overline{\chi}}'_{\mathrm{m}} \cdot \overline{\overline{k}} \cdot \overline{E} \\
&= -i\omega\varepsilon_0 \left(\overline{\overline{\chi}}_{\mathrm{e}} - \frac{1}{\omega^2 \mu_0 \varepsilon_0}\overline{\overline{k}} \cdot \overline{\overline{\chi}}'_{\mathrm{m}} \cdot \overline{\overline{k}} \right) \cdot \overline{E}
\end{aligned}$$

Thus

$$\boxed{\overline{J} = -i\omega\varepsilon_0 \overline{\overline{\chi}}_{\mathrm{e}}^{\,\mathrm{eff}} \cdot \overline{E}, \quad \text{where} \quad \overline{\overline{\chi}}_{\mathrm{e}}^{\,\mathrm{eff}} = \overline{\overline{\chi}}_{\mathrm{e}} - \frac{c^2}{\omega^2}\overline{\overline{k}} \cdot \overline{\overline{\chi}}'_{\mathrm{m}} \cdot \overline{\overline{k}}}$$

(c) We begin with the result for the constitutive relation from the previous part

$$\overline{J} = -i\omega\varepsilon_0 \left(\overline{\overline{\chi}}_{\mathrm{e}} - \frac{c^2}{\omega^2}\overline{\overline{k}} \cdot \overline{\overline{\chi}}'_{\mathrm{m}} \cdot \overline{\overline{k}} \right) \cdot \overline{E}.$$

For an electrically and magnetically isotropic medium

$$\overline{\overline{\chi}}_{\mathrm{m}} = \chi_{\mathrm{m}}\overline{\overline{I}} \quad \text{and} \quad \overline{\overline{\chi}}_{\mathrm{e}} = \chi_{\mathrm{e}}\overline{\overline{I}}.$$

So, we can write

$$\begin{aligned}
\overline{\overline{\chi}}'_{\mathrm{m}} &= \overline{\overline{\chi}}_{\mathrm{m}} \cdot \left[\overline{\overline{I}} + \overline{\overline{\chi}}_{\mathrm{m}} \right]^{-1} \\
&= \chi_{\mathrm{m}} (1 + \chi_{\mathrm{m}})^{-1} \overline{\overline{I}} \\
&= \frac{\chi_{\mathrm{m}}}{1 + \chi_{\mathrm{m}}}\overline{\overline{I}} \\
&= \chi'_{\mathrm{m}}\overline{\overline{I}}
\end{aligned}$$

The constitutive relation becomes

$$\overline{J} = -i\omega\varepsilon_0 \left(\chi_e \overline{\overline{I}} - \chi_m' \frac{c^2}{\omega^2} \overline{\overline{k}} \cdot \overline{\overline{k}} \right) \cdot \overline{E}$$

But, from Maxwell's equations, (8.46) and (8.47),

$$\overline{\overline{k}} \cdot \overline{E} = \omega \overline{B}$$

$$\overline{\overline{k}} \cdot \overline{B} = -\frac{\omega}{c^2} \overline{E} - i\mu_0 \overline{J}.$$

Eliminating $\overline{B}$ from the second equation gives

$$\overline{\overline{k}} \cdot \overline{\overline{k}} \cdot \overline{E} = -\frac{\omega^2}{c^2} \overline{E} - i\omega\mu_0 \overline{J}$$

Thus

$$\overline{J} = -i\omega\varepsilon_0 \left(\chi_e \overline{E} - \chi_m' \frac{c^2}{\omega^2} \left(\frac{-\omega^2}{c^2} \overline{E} - i\omega\mu_0 \overline{J} \right) \right)$$

$$= -i\omega\varepsilon_0 \left(\chi_e + \chi_m' \right) \overline{E} + \omega^2 \mu_0 \varepsilon_0 \chi_m' \frac{c^2}{\omega^2} \overline{J}$$

$$= -i\omega\varepsilon_0 \left(\chi_e + \chi_m' \right) \overline{E} + \chi_m' \overline{J}$$

Solving for $\overline{J}$ gives

$$\overline{J} \left(1 - \chi_m' \right) = -i\omega\varepsilon_0 \left(\chi_e + \chi_m' \right) \overline{E}$$

Thus

$$\boxed{ \therefore \overline{J} = -i\omega\varepsilon_0 \frac{\left(\chi_e + \chi_m' \right)}{\left(1 - \chi_m' \right)} \overline{E} \ \ \text{or} \ \ \overline{J} = -i\omega\varepsilon_0 \chi_e^{\text{eff}} \overline{E}. }$$

Problem 8.8 Prove the vector identity (8.39)

$$(\mathbf{M} \times \mathbf{V}) \times \mathbf{r} = -2\mathbf{M}. \tag{8.60}$$

This identity is used in the derivation of the expression for the magnetic moment (8.40).

Solution 8.8

We expand the vector as

$$\overline{M} \times \nabla = \hat{x}(M_y \partial_z - M_z \partial_y) + \hat{y}(M_z \partial_x - M_x \partial_z) + \hat{z}(M_x \partial_y - M_y \partial_x)$$

where

$$\partial_x = \frac{\partial}{\partial x}, \quad \partial_y = \frac{\partial}{\partial y} \quad \text{and} \quad \partial_z = \frac{\partial}{\partial z}.$$

Then

$$
\begin{aligned}
(\overline{M} \times \nabla) \times \bar{r} = \hat{x} \left[(M_z \partial_x - M_x \partial_z) z - (M_x \partial_y - M_y \partial_x) y \right] \\
+ \hat{y} \left[(M_x \partial_y - M_y \partial_x) x - (M_y \partial_z - M_z \partial_y) z \right] \\
+ \hat{z} \left[(M_y \partial_z - M_z \partial_y) y - (M_z \partial_x - M_x \partial_z) x \right]
\end{aligned}
$$

Therefore

$$(\overline{M} \times \nabla) \times \bar{r} = \hat{x}(-M_x - M_x) + \hat{y}(-M_y - M_y) + \hat{z}(-M_z - M_z),$$

hence proving that

$$\boxed{(\overline{M} \times \nabla) \times \bar{r} = -2\overline{M}.}$$

Problem 8.9 Consider a sphere of a magnetic insulating material with radius R. The material is electrically isotropic with susceptibility $\chi_e = 3$. The time-varying magnetization is directed along the z axis and is of the form

$$\mathbf{M} = M_0 e^{-i\omega t} \hat{\mathbf{z}}. \tag{8.61}$$

(a) Assuming the magnetostatic approximation is valid, use Faraday's law

$$\nabla \times \mathbf{E} = -\frac{\partial \mathbf{B}}{\partial t}, \tag{8.62}$$

to find the first order electric field. Show that the oscillating magnetization induces an electric field given approximately by

$$E_\phi = \frac{1}{3} i \omega \mu_0 M_0 r \tag{8.63}$$

where r is the radial cylindrical coordinate.

(b) Using the field found in part (a), evaluate the first integral on the right-hand side of the magnetic moment expression (8.40)

$$\mu = \frac{1}{2} \int_V \mathbf{r} \times \frac{\partial \mathbf{P}}{\partial t} dv + \int_V \mathbf{M} dv. \tag{8.64}$$

Show that it is negligible compared to the second term only if $R \ll \lambda_0$, where λ_0 is the wavelength in free space.

Solution 8.9

(a) In the magnetostatic approximation, we can use the demagnetizing factor to obtain $\overline{H}$ from $\overline{M}$. For a sphere, the demagnetizing factor is $1/3$ (see Sect. 5.2), So $\overline{H} = -\overline{M}/3$. Thus

$$\overline{B} = \mu_0 \left(\overline{H} + \overline{M} \right)$$
$$= \mu_0 \frac{2}{3} \overline{M}.$$

We can use Faraday's law to obtain an estimate of the first order electric field from the magnetostatic approximation for $\overline{B}$

$$\nabla \times \overline{E} = -\frac{\partial \overline{B}}{\partial t} = i\omega\mu_0 \frac{2}{3} \overline{M},$$

where $\overline{M} = M_0 e^{-i\omega t} \hat{z}$. Using cyclindrical coordinates,

$$\left(\nabla \times \overline{E} \right)_z = i\omega\mu_0 \frac{2}{3} M_0$$

$$\frac{1}{r} \left[\frac{\partial \left(r E_\phi \right)}{\partial r} - \frac{\cancel{\partial E_r}}{\cancel{\partial \phi}}^{0} \right] = i\omega\mu_0 \frac{2}{3} M_0.$$

Since the source $\overline{M}$ does not depend on ϕ, we conclude from the symmetry of the problem that the second term on the left hand side is zero. We can now integrate both sides to obtain E_ϕ

$$r E_\phi = i\omega\mu_0 \frac{2}{3} M_0 \int r dr$$
$$= i\omega\mu_0 \frac{2}{3} M_0 \frac{r^2}{2}.$$

$$\boxed{\therefore \ E_\phi = i\omega\mu_0 \frac{M_0}{3} r}$$

Here we have set the constant of integration to zero since E_ϕ must vanish at the origin.

(b) Consider (8.40) for the magnetic moment of a material sample

$$\overline{\mu} = \frac{1}{2} \int_V \overline{r} \times \frac{\partial \overline{P}}{\partial t} dv + \int_V \overline{M} dv$$

where $\overline{P} = \varepsilon_0 \chi_e \overline{E}$, and $\overline{E}$ is obtained from part (a). Our goal is to determine the conditions under which the first integral can be neglected.

In cylindrical coordinates, the volume integration over a sphere is

$$\int_V dv = 2 \int_0^R dz \int_0^{\sqrt{R^2 - z^2}} r\,dr \int_0^{2\pi} d\phi.$$

The factor of 2 is needed since the integrals actually only cover the upper hemisphere. Carrying out the indicated integrations

$$\frac{1}{2} \int_V \overline{r} \times \frac{\partial \overline{P}}{\partial t} dv = \int_0^R dz \int_0^{\sqrt{R^2 - z^2}} r\,dr \int_0^{2\pi} d\phi\, \hat{r} \times \hat{\phi} \left(-i\omega r \varepsilon_0 \chi_e E_\phi \right)$$

$$= \int_0^R dz \int_0^{\sqrt{R^2 - z^2}} r\,dr \int_0^{2\pi} d\phi\, \hat{z} \left(-i\omega r \varepsilon_0 \chi_e i\omega \mu_0 \frac{M_0 r}{3} \right)$$

$$= \hat{z} \frac{1}{3} \omega^2 \mu_0 \varepsilon_0 \chi_e M_0 \int_0^R dz \int_0^{\sqrt{R^2 - z^2}} r^3\,dr \int_0^{2\pi} d\phi$$

$$= \hat{z} \frac{2\pi}{3} \omega^2 \mu_0 \varepsilon_0 \chi_e M_0 \int_0^R dz \frac{r^4}{4} \Big|_0^{\sqrt{R^2 - z^2}}$$

$$= \hat{z} \frac{\pi}{6} \omega^2 \mu_0 \varepsilon_0 \chi_e M_0 \int_0^R dz \left(R^2 - z^2 \right)^2$$

$$= \hat{z} \frac{\pi}{6} \omega^2 \mu_0 \varepsilon_0 \chi_e M_0 \int_0^R dz \left(R^4 - 2R^2 z^2 + z^4 \right)$$

$$= \hat{z} \frac{\pi}{6} \omega^2 \mu_0 \varepsilon_0 \chi_e M_0 \left[R^4 z - \frac{2}{3} R^2 z^3 + \frac{z^5}{5} \right]_0^R$$

$$= \hat{z} \frac{\pi}{6} \omega^2 \mu_0 \varepsilon_0 \chi_e M_0 \left[R^5 - \frac{2}{3} R^5 + \frac{R^5}{5} \right]$$

$$= \frac{\pi}{6} \omega^2 \mu_0 \varepsilon_0 \chi_e M_0 R^5 \left[\frac{15 - 10 + 3}{15} \right]$$

$$\therefore \quad \frac{1}{2} \int_V \overline{r} \times \frac{\partial \overline{P}}{\partial t} dv = \hat{z} \frac{4}{15} \frac{\pi}{3} \omega^2 \mu_0 \varepsilon_0 \chi_e M_0 R^5$$

Clearly, since $\overline{M}$ is uniform

$$\int_v \overline{M}\,dv = \hat{z}M_0\frac{4}{3}\pi R^3.$$

The condition for neglecting the first integral is, therefore

$$\frac{4}{15}\frac{\pi}{3}\omega^2\mu_0\varepsilon_0\chi_e M_0 R^5 \ll M_0\frac{4}{3}\pi R^3$$

$$\therefore \quad \frac{1}{15}\omega^2\mu_0\varepsilon_0\chi_e R^2 \ll 1$$

But

$$\omega^2\mu_0\varepsilon_0 = k_0^2 = \left(\frac{2\pi}{\lambda_0}\right)^2,$$

where λ_0 is the wavelength in free space. Hence

$$\left(\frac{2\pi}{\lambda_0}\right)^2\chi_e R^2 \ll 15$$

$$R^2 \ll \frac{15}{(2\pi)^2}\frac{\lambda_0^2}{\chi_e}$$

$$\text{or, } R \ll \sqrt{\frac{15}{\chi_e}\frac{\lambda_0}{2\pi}}$$

For $\chi_e = 3$,

$$\boxed{R \ll 0.36\lambda_0}$$

Problem 8.10   Consider light propagation in a medium magnetized along the z direction. The permittivity is of the form

$$\overline{\varepsilon} = \varepsilon_0\begin{bmatrix} \varepsilon_r & -i\varepsilon_g & 0 \\ i\varepsilon_g & \varepsilon_r & 0 \\ 0 & 0 & \varepsilon_r \end{bmatrix}. \tag{8.65}$$

(a) Assuming the permeability is that of free space, evaluate the determinant of the dispersion matrix (see (4.32))

$$\left|\mathbf{k}\cdot\overline{\mu}^{-1}\cdot\mathbf{k} + \omega^2\overline{\varepsilon}\right| = 0 \tag{8.66}$$

for $\mathbf{k} = k\hat{z}$ and show that the the normal mode dispersion relations are

$$k_\pm^2 = \omega^2\mu_0\varepsilon_0(\varepsilon_r \pm \varepsilon_g). \tag{8.67}$$

(b) The Faraday rotation per unit length is given by (see Problem 4.4)

$$\phi_F = \frac{1}{2}(k_+ - k_-). \tag{8.68}$$

For $\varepsilon_g = f M_S$, find an approximation to the Faraday rotation per unit length in the limit $\varepsilon_g \ll \varepsilon_r$ and show that

$$\phi_F = \frac{k_0 M_S f}{2\sqrt{\varepsilon_r}}. \tag{8.69}$$

Solution 8.10

(a) If $\overline{\overline{\mu}} = \mu_0 \overline{\overline{I}}$, then $\left| \overline{k} \cdot \overline{\overline{\mu}}^{-1} \cdot \overline{k} + \omega^2 \overline{\overline{\varepsilon}} \right| = 0$ reduces to

$$\left| \overline{k} \cdot \overline{k} + \omega^2 \mu_0 \overline{\overline{\varepsilon}} \right| = 0$$

$$\text{or,} \quad \left| \overline{k}\,\overline{k} - k^2 \overline{\overline{I}} + \omega^2 \mu_0 \overline{\overline{\varepsilon}} \right| = 0.$$

Using (8.65) for the permittivity and expanding into matrix notation

$$\det \begin{bmatrix} k_0^2 \varepsilon_r - k^2 & -ik_0^2 \varepsilon_g & 0 \\ ik_0^2 \varepsilon_g & k_0^2 \varepsilon_r - k^2 & 0 \\ 0 & 0 & k_0^2 \varepsilon_r \end{bmatrix} = 0$$

we obtain

$$k_0^2 \varepsilon_r \left[\left[k_0^2 \varepsilon_r - k^2 \right]^2 - k_0^4 \varepsilon_g^2 \right] = 0$$

$$k_0^2 \varepsilon_r - k^2 = \pm k_0^2 \varepsilon_g$$

$$k^2 = k_0^2 \left[\varepsilon_r \mp \varepsilon_g \right]$$

or

$$\boxed{k_\pm^2 = \omega^2 \mu_0 \varepsilon_0 \left[\varepsilon_r \pm \varepsilon_g \right].}$$

(b) The Faraday rotation per unit length can now be written as

$$\phi_F = \frac{1}{2}(k_+ - k_-)$$

$$= \frac{1}{2}k_0 \left[\sqrt{\varepsilon_r + \varepsilon_g} - \sqrt{\varepsilon_r - \varepsilon_g} \right] = \frac{k_0\sqrt{\varepsilon_r}}{2} \left[\sqrt{1 + \frac{\varepsilon_g}{\varepsilon_r}} - \sqrt{1 - \frac{\varepsilon_g}{\varepsilon_r}} \right].$$

For $\varepsilon_g \ll \varepsilon_r$

$$\sqrt{1 \pm \frac{\varepsilon_g}{\varepsilon_r}} \approx 1 \pm \frac{\varepsilon_g}{2\varepsilon_r}.$$

Hence

$$
\begin{aligned}
\phi_{\mathrm{F}} &\approx \frac{k_0 \sqrt{\varepsilon_r}}{2}\left[\sqrt{1 + \frac{\varepsilon_g}{2\varepsilon_r}} - \sqrt{1 + \frac{\varepsilon_g}{2\varepsilon_r}}\right] \\
&= \frac{k_0 \varepsilon_g}{2\sqrt{\varepsilon_r}}.
\end{aligned}
$$

For $\varepsilon_g = f M_S$, we recover

$$
\boxed{\phi_{\mathrm{F}} = \frac{k_0 f M_S}{2\sqrt{\varepsilon_r}}.}
$$

8.3 Coupled-Mode Theory

The TE_n and TM_n modes of dielectric waveguides that we have analyzed are orthogonal if the medium is uniform and non-time-varying. In other words, if we launch a TE_n mode, it will not couple to or "leak" into the TM_n mode as it propagates. In contrast, if the permittivity (or permeability) of the medium has small variations in space and time owing to the presence of propagating spin waves, this may no longer be the case. In particular, if specific conditions are satisfied, the energy will oscillate back and forth between the two modes as they propagate, with the spatial period of the oscillation dependent on the strength of the perturbation or coupling.

To understand the required conditions for this coupling to occur, it is useful to think in terms of photons and magnons, i.e., wave packets characterized by the energy $\hbar\omega$ and the momentum $\hbar k$. Since the TE_n and TM_n modes have different dispersion relations, if they have the same frequency (energy), they will not have the same momentum (k vector). Consequently, if a TE_n mode converted to a TM_n mode, conservation of momentum would be violated. This is where the magnons come in. We can have mode conversion resulting from scattering from magnons if we satisfy energy *and* momentum conservation, i.e., if

$$
\omega_{\mathrm{TE}} = \omega_{\mathrm{TM}} \pm \omega_{\mathrm{SW}}, \tag{8.70}
$$

and

$$
\mathbf{k}_{\mathrm{TE}}^{(n)} = \mathbf{k}_{\mathrm{TM}}^{(m)} \pm \mathbf{k}_{\mathrm{SW}}. \tag{8.71}
$$

We can model this mode conversion by taking the amplitude of the modes to be functions of position that are slowly-varying on the scale of an optical wavelength. For example, if we take the amplitude of the two modes as $A(z)$ and $B(z)$ (it doesn't really matter which mode we assign to A which we assign to B), the mode amplitudes will vary with z according to the *coupled-mode equations*

$$\frac{\partial A}{\partial z} = \kappa_{ab} B e^{-i\Delta z}, \tag{8.72a}$$

$$\frac{\partial B}{\partial z} = \kappa_{ba} A e^{+i\Delta z}, \tag{8.72b}$$

where

$$\kappa_{ab} = -\kappa_{ba}^* \triangleq \kappa \tag{8.73}$$

is the coupling coefficient and Δ is the phase mismatch

$$\Delta = k_A - k_B. \tag{8.74}$$

The magnitude of the coupling coefficient, $|\kappa|$, for the case of collinear spin wave optical interactions is discussed in Sect. 8.4.

In Problem 8.11, we derive the general solutions to (8.72) and demonstrate the periodic transfer of power between the two coupled modes. When there is complete phase matching, $\Delta = 0$, and power is transferred completely from mode A to mode B over a characteristic length

$$L_c = \frac{\pi}{2|\kappa|}. \tag{8.75}$$

The relationship (8.73) is a consequence of imposing the requirement of energy conservation as the modes propagate. Specifically, the total energy in both modes is proportional to $|A(z)|^2 + |B(z)|^2$. If the energy remains constant, then the derivative of this quantity with respect to z must be zero. Imposing this condition leads to (8.73), as shown in Problem 8.12.

Inherent in our treatment is the assumption that the modal solutions $A(z)$ and $B(z)$ are unperturbed by the coupling between them. When one assumes that the excitation of a coupled system is given by a linear superposition of the modes of an uncoupled system, we are using field patterns that have been obtained in the absence of coupling. A more accurate description for the propagation of modes in a coupled system is obtained by defining supermodes [2, Sect. 22.9], or by using a variational principle [5].

Problem 8.11 Derive the general solutions

$$A(z) = e^{-i\Delta z/2} \left[A(0) \cos \beta_0 z + \frac{\kappa_{ab} B(0) + (i\Delta/2)A(0)}{\beta_0} \sin \beta_0 z \right], \tag{8.76a}$$

$$B(z) = e^{i\Delta z/2} \left[B(0) \cos \beta_0 z + \frac{\kappa_{ba} A(0) - (i\Delta/2)B(0)}{\beta_0} \sin \beta_0 z \right], \tag{8.76b}$$

to the coupled-mode equations (8.72) where $\beta_0^2 = |\kappa_{ab}|^2 + |\Delta/2|^2$. The following steps will assist you:

(a) Obtain two boundary condition equations by imposing the initial values $A(0)$ and $B(0)$ on the trial solutions

$$A = \left[C_1 e^{i\beta_0 z} + C_2 e^{-i\beta_0 z}\right] e^{-i\Delta z/2}, \tag{8.77a}$$

$$B = \left[D_1 e^{i\beta_0 z} + D_2 e^{-i\beta_0 z}\right] e^{i\Delta z/2}. \tag{8.77b}$$

(b) Substitute the trial functions (8.77) into the coupled-mode equations (8.76) to obtain

$$e^{i\beta_0 z} \left[(i\beta_0 - i\Delta/2)C_1 - \kappa_{ab} D_1\right] - e^{-i\beta_0 z} \left[(i\beta_0 + i\Delta/2)C_2 + \kappa_{ab} D_2\right] = 0, \tag{8.78a}$$

$$e^{i\beta_0 z} \left[(i\beta_0 + i\Delta/2)D_1 - \kappa_{ba} C_1\right] - e^{-i\beta_0 z} \left[(i\beta_0 - i\Delta/2)D_2 + \kappa_{ba} C_2\right] = 0. \tag{8.78b}$$

Using (8.78a), find expressions for the ratios D_1/C_1 and D_2/C_2. Use these ratios to eliminate D_1 and D_2 from the boundary condition equations obtained in part (a). The result is a set of two simultaneous equations for C_1 and C_2.

(c) Solve the equations for C_1 and C_2 obtained in part (b). Substitute the result into the trial solution (8.77a), group terms, and simplify to obtain the general solution (8.76a).

(d) Next, find expressions for the ratios C_1/D_1 and C_2/D_2 using (8.78b). Use these ratios to eliminate C_1 and C_2 from the boundary condition equations obtained in part (a). The result is a set of two simultaneous equations for D_1 and D_2.

(e) Solve the equations obtained in part (d) for D_1 and D_2. Substitute the result into the trial solution (8.77b), group terms, and simplify to obtain the general solution (8.76b).

(f) Consider the special case when power is launched into one of the modes, i.e., $A(0) = 1$ and $B(0) = 0$. Plot the mode power as a function of propagation distance for the two cases

   (i) Perfect phase matching, $\Delta = 0$,
   (ii) With a phase mismatch, $\Delta/\kappa = 3$.

Solution 8.11

Consider the coupled-mode equations (8.76)

$$\frac{\partial A}{\partial z} = \kappa_{ab} B e^{-i\Delta z},$$

$$\frac{\partial B}{\partial z} = \kappa_{ba} A e^{+i\Delta z}.$$

(a) To find the general solution, consider the trial functions (8.77)

$$A = \left[C_1 e^{i\beta_0 z} + C_2 e^{-i\beta_0 z} \right] e^{-i\Delta z/2},$$

$$B = \left[D_1 e^{i\beta_0 z} + D_2 e^{-i\beta_0 z} \right] e^{i\Delta z/2}.$$

The initial values at $z = 0$ are

$$\boxed{A\,(0) = C_1 + C_2, \quad B\,(0) = D_1 + D_2.}$$

(b) Substituting the trial solutions into the coupled-mode equations gives (8.78a) and (8.78b)

$$e^{i\beta_0 z}\left[i\left(\beta_0 - \frac{\Delta}{2}\right)C_1 - \kappa_{ab}D_1 \right] - e^{-i\beta_0 z}\left[i\left(\beta_0 + \frac{\Delta}{2}\right)C_2 + \kappa_{ab}D_2 \right] = 0,$$

$$e^{i\beta_0 z}\left[i\left(\beta_0 + \frac{\Delta}{2}\right)D_1 - \kappa_{ba}C_1 \right] - e^{-i\beta_0 z}\left[i\left(\beta_0 - \frac{\Delta}{2}\right)D_2 + \kappa_{ba}C_2 \right] = 0.$$

For this equation to be satisfied for all values of z, the coefficient of each exponential must vanish. First consider (8.78a). Setting the coefficient of $e^{i\beta_0 z}$ to zero gives

$$i\left(\beta_0 - \frac{\Delta}{2}\right)C_1 = \kappa_{ab}D_1.$$

Similarly, from the co-efficient of $e^{-i\beta_0 z}$ we get

$$i\left(\beta_0 + \frac{\Delta}{2}\right)C_2 = -\kappa_{ab}D_2.$$

From part (a) we have

$$B(0) = D_1 + D_2$$

Using the above equations to eliminate D_1 and D_2 gives

$$B\,(0) = \frac{i}{\kappa_{ab}}\left(\beta_0 - \frac{\Delta}{2}\right)C_1 - \frac{i}{\kappa_{ab}}\left(\beta_0 + \frac{\Delta}{2}\right)C_2.$$

Also from part (a)

$$A\,(0) = C_1 + C_2.$$

Thus, we have two simultaneous equations in C_1 and C_2

$$\boxed{\begin{bmatrix} 1 & 1 \\ \frac{i}{\kappa_{ab}}(\beta_0 - \Delta/2) & -\frac{i}{\kappa_{ab}}(\beta_0 + \Delta/2) \end{bmatrix}\begin{bmatrix} C_1 \\ C_2 \end{bmatrix} = \begin{bmatrix} A(0) \\ B(0) \end{bmatrix}}$$

(c) To solve for C_1 and C_2, multiply from the left by the inverse of the square matrix

$$\begin{bmatrix} C_1 \\ C_2 \end{bmatrix} = \frac{1}{-\frac{i}{\kappa_{ab}}(\beta_0 + \Delta/2) - \frac{i}{\kappa_{ab}}(\beta_0 - \Delta/2)}$$

$$\times \begin{bmatrix} -\frac{i}{\kappa_{qb}}(\beta_0 + \Delta/2) & -1 \\ -\frac{i}{\kappa_{ab}}(\beta_0 - \Delta/2) & 1 \end{bmatrix} \begin{bmatrix} A(0) \\ B(0) \end{bmatrix}$$

$$\therefore \quad \begin{bmatrix} C_1 \\ C_2 \end{bmatrix} = \frac{1}{2\beta_0} \begin{bmatrix} \beta_0 + \Delta/2 & -i\kappa_{ab} \\ \beta_0 - \Delta/2 & i\kappa_{ab} \end{bmatrix} \begin{bmatrix} A(0) \\ B(0) \end{bmatrix}$$

$$\Rightarrow \quad C_1 = \frac{1}{2\beta_0} [(\beta_0 + \Delta/2) A(0) - i\kappa_{ab} B(0)]$$

$$C_2 = \frac{1}{2\beta_0} [(\beta_0 - \Delta/2) A(0) + i\kappa_{ab} B(0)]$$

Recall the trial solution

$$A = \left[C_1 e^{i\beta_0 z} + C_2 e^{-i\beta_0 z} \right] e^{-i\Delta z/2}.$$

Substituting for C_1 and C_2 gives

$$A e^{i\Delta z/2} = \frac{1}{2\beta_0} \left[\left(\beta_0 + \frac{\Delta}{2} \right) A(0) - i\kappa_{ab} B(0) \right] e^{i\beta_0 z}$$

$$+ \frac{1}{2\beta_0} \left[\left(\beta_0 - \frac{\Delta}{2} \right) A(0) + i\kappa_{ab} B(0) \right] e^{-i\beta_0 z}$$

$$= A(0) \left[\frac{e^{i\beta_0 z} + e^{-i\beta_0 z}}{2} \right] + i\frac{A(0)}{\beta_0}\frac{\Delta}{2} \left[\frac{e^{i\beta_0 z} - e^{-i\beta_0 z}}{2i} \right]$$

$$+ \frac{B(0)\kappa_{ab}}{\beta_0} \left[\frac{e^{i\beta_0 z} - e^{-i\beta_0 z}}{2i} \right]$$

$$= A(0) \cos \beta_0 z + \frac{i A(0)\Delta}{2\beta_0} \sin \beta_0 z + \frac{B(0)\kappa_{ab}}{\beta_0} \sin \beta_0 z$$

Finally

$$\boxed{A(z) = e^{-i\Delta z/2} \left[A(0) \cos \beta_0 z + \frac{\kappa_{ab} B(0) + \left(i\frac{\Delta}{2} \right) A(0)}{\beta_0} \sin \beta_0 z \right]}$$

(d) Setting the coefficient of the exponentials in (8.78b) to zero gives

$$i\left(\beta_0 + \frac{\Delta}{2} \right) D_1 = \kappa_{ba} C_1,$$

$$i\left(\beta_0 - \frac{\Delta}{2} \right) D_2 = -\kappa_{ba} C_2.$$

Also from part (a)

$$A(0) = C_1 + C_2$$

$$= \frac{i}{\kappa_{ba}} \left(\beta_0 + \frac{\Delta}{2} \right) D_1 - \frac{i}{\kappa_{ba}} \left(\beta_0 - \frac{\Delta}{2} \right) D_2$$

Along with $B(0) = D_1 + D_2$ this gives the simultaneous equations

$$\begin{bmatrix} \frac{i}{\kappa_{ba}}(\beta_0 + \Delta/2) & -\frac{i}{\kappa_{ba}}(\beta_0 - \Delta/2) \\ 1 & 1 \end{bmatrix} \begin{bmatrix} D_1 \\ D_2 \end{bmatrix} = \begin{bmatrix} A(0) \\ B(0) \end{bmatrix}$$

(e) Multiplying from the left by the inverse matrix as before gives

$$\begin{bmatrix} D_1 \\ D_2 \end{bmatrix} = \frac{1}{2\beta_0} \begin{bmatrix} -i\kappa_{ba} & \beta_0 - \Delta/2 \\ i\kappa_{ba} & \beta_0 + \Delta/2 \end{bmatrix} \begin{bmatrix} A(0) \\ B(0) \end{bmatrix}$$

or

$$D_1 = \frac{1}{2\beta_0} \left[-i\kappa_{ba} A(0) + \left(\beta_0 - \frac{\Delta}{2} \right) B(0) \right],$$

$$D_2 = \frac{1}{2\beta_0} \left[i\kappa_{ba} A(0) + \left(\beta_0 + \frac{\Delta}{2} \right) B(0) \right].$$

Substituting this into the trial solution (8.77b) gives

$$e^{-i\Delta z/2} B = \frac{e^{i\beta_0 z}}{2\beta_0} \left[-i\kappa_{ba} A(0) + \left(\beta_0 - \frac{\Delta}{2} \right) B(0) \right]$$

$$+ \frac{e^{-i\beta_0 z}}{2\beta_0} \left[i\kappa_{ba} A(0) + \left(\beta_0 + \frac{\Delta}{2} \right) B(0) \right]$$

Collecting terms and simplifying

$$e^{-i\Delta z/2} B = \frac{A(0)}{\beta_0} \kappa_{ba} \left[\frac{e^{i\beta_0 z} - e^{-i\beta_0 z}}{2i} \right] + B(0) \left[\frac{e^{i\beta_0 z} + e^{-i\beta_0 z}}{2} \right]$$

$$- i \frac{B(0)(\Delta/2)}{\beta_0} \left[\frac{e^{i\beta_0 z} - e^{-i\beta_0 z}}{2i} \right]$$

and finally

$$B = e^{i\Delta z/2} \left[B(0) \cos \beta_0 z + \frac{\kappa_{ba} A(0) - (i\Delta/2) B(0)}{\beta_0} \sin \beta_0 z \right]$$

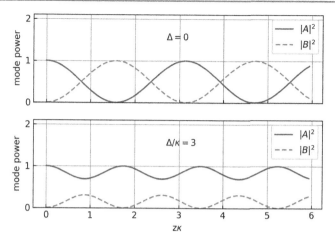

Fig. 8.3 Mode power as a function of propagation distance z for $\Delta = 0$ and $\Delta/\kappa = 3$. (Adapted from [4] with permission.)

(f) When we use the conditions $A(0) = 1$ and $B(0) = 0$, (8.76) reduces to

$$|A|^2 = \cos^2\left[z\sqrt{|\kappa^2| + (\Delta/2)^2}\right] + \frac{(\Delta/2)^2}{|\kappa|^2 + (\Delta/2)^2} \sin^2\left[z\sqrt{|\kappa^2| + (\Delta/2)^2}\right],$$

$$(8.79a)$$

$$|B|^2 = \frac{|\kappa|^2}{|\kappa|^2 + (\Delta/2)^2} \sin^2\left[z\sqrt{|\kappa^2| + (\Delta/2)^2}\right].$$

$$(8.79b)$$

The Python code for plotting these equations is given below, and the resulting plots are shown in Fig. 8.3 for the two cases of $\Delta = 0$, and $\Delta/\kappa = 3$. For $\Delta = 0$, the converted mode amplitude is simply $B = \sin(|\kappa|z)$, and we observe a complete transfer of power from A to B over the coupling length $L_c = \pi/2|\kappa|$, or when $z\kappa = \pi/2$. In the case of a partial phase match, $\Delta = 3\kappa$, we observe an incomplete transfer of power and a shift in the coupling length.

```python
from matplotlib import pyplot as plt
import numpy as np

# calculate optical mode conversion using coupled mode theory

# define vectorized functions

def mycos(x):
  return np.cos(x)
def mysin(x):
  return np.sin(x)
```

```python
mycosv = np.vectorize(mycos)
mysinv = np.vectorize(mysin)

# Set values of Delta/Kappa
DK1 = 0;
DK2 = 3;

# number of points N to plot, and maximum value of zk to plot
N = 100
zkmax = 6

# create zk arrays

zk = np.arange(0,zkmax,zkmax/(N-1))
c1 = np.sqrt(1+(DK1/2)**2)
c2 = np.sqrt(1+(DK2/2)**2)
zk1 = zk*c1
zk2 = zk*c2

# calculate cosine and sine terms
C1 = mycosv(zk1)**2
S1 = mysinv(zk1)**2
C2 = mycosv(zk2)**2
S2 = mysinv(zk2)**2

# calculate solutions for DK=0

AA1 = C1
BB1 = S1

# calculate solutions for DK = 3

AA2 = C2 + (DK2/2)**2*S2/c2**2
BB2 = S2/c2**2

# plot and download

fig, ax = plt.subplots(2, sharex=True, sharey=True)
ax[0].plot(zk, AA1,label='$|A|^2$')
ax[0].plot(zk, BB1,linestyle='--',label='$|B|^2$')
ax[0].legend(loc='upper right')
ax[0].set_ylim(0,2.1)
ax[0].text(2.7,1.5,'$\Delta=0$')
```

```
ax[1].plot(zk, AA2,label='$|A|^2$')
ax[1].plot(zk, BB2,linestyle='--',label='$|B|^2$')
ax[1].legend(loc='upper right')
ax[1].set_ylim(0,2.1)
ax[1].text(2.7,1.5,'$\Delta/\kappa=3$')

for ax1 in ax:
    ax1.tick_params(direction='in')
    ax1.grid(True)

#fig, ax=plt.subplots()
#ax.plot(tn,C,tn,Gn,tn,P)
ax[1].set(xlabel='z$\kappa$')
ax[0].set(ylabel='mode power')
ax[1].set(ylabel='mode power')

fig1 = plt.gcf()
plt.show()
```

Problem 8.12 Verify that the general coupled-mode solutions (8.76a) and (8.76b) satisfy the energy conservation relation

$$\frac{d}{dz}\left(|A|^2 + |B|^2\right) = 0. \tag{8.80}$$

Solution 8.12

To demonstrate that energy conservation, in coupled-mode theory, satisfies

$$\frac{d}{dz}\left[|A|^2 + |B|^2\right] = 0,$$

we begin with the solutions (8.76a) and (8.76b)

$$A = e^{-i\Delta z/2}\left[A(0)\cos\beta_0 z + \frac{\kappa_{ab}B(0) + (i\Delta/2)A(0)}{\beta_0}\sin\beta_0 z\right],$$

$$B = e^{i\Delta z/2}\left[B(0)\cos\beta_0 z - \frac{\kappa_{ba}A(0) - (i\Delta/2)B(0)}{\beta_0}\sin\beta_0 z\right].$$

Taking the magnitude squared

$$|A|^2 = AA^*$$

$$= \left[A(0) \cos \beta_0 z + \frac{\kappa_{ab} B(0) + (i\Delta/2) A(0)}{\beta_0} \sin \beta_0 z \right]$$

$$\times \left[A^*(0) \cos \beta_0 z + \frac{\kappa_{ab}^* B^*(0) - (i\Delta/2) A^*(0)}{\beta_0} \sin \beta_0 z \right]$$

$$= |A(0)|^2 \cos^2 \beta_0 z + \left[\frac{\kappa_{ab} B(0) + (i\Delta/2) A(0)}{\beta_0} \right] \sin \beta_0 z \, A^*(0) \cos \beta_0 z$$

$$+ \left[\frac{\kappa_{ab}^* B^*(0) - (i\Delta/2) A^*(0)}{\beta_0} \right] A(0) \cos \beta_0 z \sin \beta_0 z + \left[|\kappa_{ab}|^2 |B(0)|^2 \right.$$

$$+ i\frac{\Delta}{2} A(0) \kappa_{ab}^* B^*(0) - i\frac{\Delta}{2} A^*(0) B(0) \kappa_{ab} + \left(\frac{\Delta}{2} \right)^2 |A(0)|^2 \left] \frac{\sin^2 \beta_0 z}{\beta_0^2} \right.$$

or,

$$|A|^2 = |A(0)|^2 \cos^2 \beta_0 z + \left[\kappa_{ab} B(0) A^*(0) + \kappa_{ab}^* B^*(0) A(0) \right] \frac{\cos \beta_0 z \sin \beta_0 z}{\beta_0}$$

$$+ \left[|\kappa_{ab}|^2 |B(0)|^2 + i\frac{\Delta}{2} A(0) \kappa_{ab}^* B^*(0) - i\frac{\Delta}{2} A^*(0) B(0) \kappa_{ab} \right.$$

$$+ \left(\frac{\Delta}{2} \right)^2 |A(0)|^2 \left] \frac{\sin^2 \beta_0 z}{\beta_0^2}. \right.$$

To get $|B|^2$, we make the substitutions

$$\Delta \to -\Delta$$
$$A(0) \to B(0)$$
$$B(0) \to A(0)$$
$$\kappa_{ab} \to \kappa_{ba}$$

and get

$$|B|^2 = |B(0)|^2 \cos^2 \beta_0 z + \left[\kappa_{ba} A(0) B^*(0) + \kappa_{ba}^* A^*(0) B(0) \right] \frac{\cos \beta_0 z \sin \beta_0 z}{\beta_0}$$

$$+ \left[|\kappa_{ba}|^2 |A(0)|^2 - \frac{i\Delta}{2} B(0) \kappa_{ba}^* A^*(0) + \frac{i\Delta}{2} B^*(0) A(0) \kappa_{ba} \right.$$

$$+ \left(\frac{\Delta}{2} \right)^2 |B(0)|^2 \left] \frac{\sin^2 \beta_0 z}{\beta_0^2}. \right.$$

Adding $|A|^2$ and $|B|^2$ and assuming $\kappa_{ab} = -\kappa_{ba}^*$ gives

$$|A|^2 + |B|^2 = \left[|A(0)|^2 + |B(0)|^2\right] \cos^2 \beta_0 z$$

$$+ \left[\underbrace{\kappa_{ab} B(0) A^*(0) + \kappa_{ab}^* B^*(0) A(0) + \kappa_{ba} A(0) B^*(0) + \kappa_{ba}^* A^*(0) B(0)}_{\text{terms cancel if } \kappa_{ab} = -\kappa_{ba}^*}\right]$$

$$\times \frac{\cos \beta_0 z \sin \beta_0 z}{\beta_0}$$

$$+ \left[|\kappa|^2 \left[|A(0)|^2 + |B(0)|^2\right] + \frac{i\Delta}{2} A(0) B^*(0) \underbrace{\left[\kappa_{ba} + \kappa_{ab}^*\right]}_{\text{terms cancel}}\right.$$

$$\left. - \frac{i\Delta}{2} A^*(0) B(0) \underbrace{\left[\kappa_{ab} + \kappa_{ba}^*\right]}_{\text{terms cancel}} + \left(\frac{\Delta}{2}\right)^2 \left[|A(0)|^2 + |B(0)|^2\right]\right] \frac{\sin^2 \beta_0 z}{\beta_0^2}$$

Collecting the surviving terms gives

$$|A|^2 + |B|^2 = \left[|A(0)|^2 + |B(0)|^2\right] \cos^2 \beta_0 z$$

$$+ \underbrace{\frac{|\kappa|^2 + (\frac{\Delta}{2})^2}{\beta_0^2}}_{= 1 \text{ if } \beta_0^2 = |\kappa|^2 + (\Delta/2)^2} \left[|A(0)|^2 + |B(0)|^2\right] \sin^2 \beta_0 z$$

$$= \left[|A(0)|^2 + |B(0)|^2\right] \left[\cos^2 \beta_0 z + \sin^2 \beta_0 z\right]$$

$$= \left[|A(0)|^2 + |B(0)|^2\right] = \text{Constant} (= \text{initial total power in the modes})$$

Therefore

$$\boxed{\frac{\mathrm{d}}{\mathrm{d}z}\left[|A|^2 + |B|^2\right] = \frac{\mathrm{d}}{\mathrm{d}z}\left[|A(0)|^2 + |B(0)|^2\right] = 0.}$$

8.4   Magneto-optic Coupling

The direction of the static magnetization of a ferrite, such as YIG, influences its optical permittivity. Optical fields of interest to telecommunications, at $\lambda \sim 1310\,\mathrm{nm}$ or $1550\,\mathrm{nm}$, will oscillate at about 200 THz, and it is reasonable to consider the GHz precession of $\mathbf{M}$ to be "static" relative to the optical fields. The relative permittivity tensor in such a situation can be written as

$$\bar{\bar{\varepsilon}} = \varepsilon_0 \begin{bmatrix} \varepsilon_r & -if\,M_z & if\,M_y \\ if\,M_z & \varepsilon_r & -if\,M_x \\ -if\,M_y & if\,M_x & \varepsilon_r \end{bmatrix}. \tag{8.81}$$

where f is a proportionality constant, as described in [4, Sect. 8.2.3]. The off-diagonal elements give rise to Faraday rotation, in a manner analogous to that of a magnetically gyrotropic medium (Problem 4.5). Assuming that the magnetization is saturated and parallel to the direction of light propagation, the Faraday rotation (rotation angle per unit length of propagation) is (see Problem 8.10)

$$\phi_F = \frac{k_0 M_S f}{2\sqrt{\varepsilon_r}}, \tag{8.82}$$

where k_0 is the optical wave number in free space, and M_S is the saturation magnetization of the medium. (The Faraday rotation per unit length ϕ_F differs from the total rotation angle $\theta_F = \phi_F d$.) We typically use (8.82) to determine the phenomenological parameter f from a measurement of ϕ_F.

For the case where optical guided modes are scattered by spin waves, the permittivity will have time-varying components at GHz frequencies. But these variations are slow compared to the THz optical frequencies. Hence, we can define a coupling coefficient κ_{ab} between the TE and TM optical modes. As an example, consider the magnetization

$$\mathbf{M}(t) \approx M_S \hat{\mathbf{x}} + m_y(t)\hat{\mathbf{y}} + m_z(t)\hat{\mathbf{z}}, \tag{8.83}$$

such that we have forward volume waves co-propagating with the TE and TM optical modes along $\hat{\mathbf{z}}$ in a geometry similar to Fig. 8.2. We must modify the permittivity tensor (8.81) to reflect the small-signal magnetization components, and then estimate κ_{ab} (see Problem 8.13).

Problem 8.13 Consider coupling between optical guided modes induced by a propagating spin wave. If the optical modes are far above cutoff, the optical fields are well-confined to the magnetic film. On the other hand, if $k_{SW}d \ll 1$, then the spin wave amplitude will be approximately constant over the film thickness. Under these conditions, the coupling coefficient resulting from forward volume spin waves is approximately

$$\kappa_{ab} = -\frac{1}{2}\phi_F \frac{m_{0z}}{M_S}, \tag{8.84}$$

where ϕ_F is the Faraday rotation angle per unit length (see (8.69))

$$\phi_F = \frac{M_S f k_0}{2\sqrt{\varepsilon_r}}, \tag{8.85}$$

m_{0z} is the amplitude of the z component of the small-signal magnetization $m_z(t)$, the material is magnetized along x (perpendicular to the film), the propagation is along z, and the gyrotropic permittivity tensor is given by

$$\bar{\varepsilon} = \varepsilon_0 \begin{bmatrix} \varepsilon_r & -if m_z & if m_y \\ if m_z & \varepsilon_r & -if M_S \\ -if m_y & if M_S & \varepsilon_r \end{bmatrix}.$$

Derive (8.84) by substituting the field expressions

$$e_y^{(0)}(x) = 2\sqrt{\frac{\omega\mu_0}{k_{TE}^{(0)}d}}\cos\left(\frac{\pi x}{d}\right), \quad |x| \le d/2, \tag{8.86a}$$

$$h_y^{(0)}(x) = 2\sqrt{\frac{\omega\varepsilon_f}{k_{TM}^{(0)}d}}\cos\left(\frac{\pi x}{d}\right), \quad |x| \le d/2, \tag{8.86b}$$

into the equation for the coupling constant[2]

$$\kappa_{ab} = -\frac{k_{TM}^{(m)}f}{8\varepsilon_f}\int_{-d/2}^{d/2} dx\; e_y^{(n)}(x)h_y^{(m)}(x)m_{0z}(x) \tag{8.87}$$

and simplifying. Here $\varepsilon_f = \varepsilon_0\varepsilon_r$. Take m_{0z} to be approximately constant through the film thickness.

Solution 8.13

The spin wave optical mode coupling coefficient is given by (8.87)

$$\kappa_{ab} = -\frac{k_{TM}^{(m)}f}{8\varepsilon_f}\int_{-d/2}^{d/2} dx\; e_y^{(n)}(x)h_y^{(m)}(x)m_{0z}(x).$$

Far from cutoff, the optical mode fields can be approximated by (8.86a) and (8.86b)

$$e_y^{(0)}(x) = 2\sqrt{\frac{\omega\mu_0}{k_{TE}^{(0)}d}}\cos\frac{\pi x}{d} \quad |x| \le d/2,$$

$$h_y^{(0)}(x) = 2\sqrt{\frac{\omega\varepsilon_f}{k_{TM}^{(0)}d}}\cos\frac{\pi x}{d} \quad |x| \le d/2.$$

Substituting both into (8.87) and simplifying

$$\kappa_{ab} = -\frac{k_{TM}^{(0)}}{8\varepsilon_f}f\int_{-d/2}^{d/2}\frac{m_{0z}4\omega\sqrt{\mu_0\varepsilon_f}}{d\sqrt{k_{TE}^{(0)}k_{TM}^{(0)}}}\cos^2\frac{\pi x}{d}\,dx$$

---

[2] Haus and Huang [5] provide a review of coupled-mode theory of passive structures. This expression for κ is derived in Stancil and Prabhakar [4, Sect. 8.4.3].

$$\kappa_{ab} = -\frac{k_{\mathrm{TM}}^{(0)}}{2\varepsilon_r d} f \int_{-d/2}^{d/2} \frac{m_{0z}\omega\sqrt{\mu_0\varepsilon_f}}{\sqrt{k_{\mathrm{TE}}^{(0)} k_{\mathrm{TM}}^{(0)}}} \frac{1}{2}\left(1 + \cos\frac{2\pi x}{d}\right) dx$$

$$= -\frac{k_{\mathrm{TM}}^{(0)} f}{4\varepsilon_r d} \frac{m_{0z}\omega\sqrt{\mu_0\varepsilon_f}\, d}{\sqrt{k_{\mathrm{TE}}^{(0)} k_{\mathrm{TM}}^{(0)}}}$$

Assuming $k_{\mathrm{TM}} \approx k_{\mathrm{TE}} = \omega\sqrt{\mu_0\varepsilon_f} = k_0\sqrt{\varepsilon_r}$ since the modes are tightly bound to the film,

$$\kappa_{ab} = -\frac{m_{0z} f k_0}{4\sqrt{\varepsilon_r}} = -\frac{m_{0z}}{2M_S} \frac{M_S f k_0}{2\sqrt{\varepsilon_r}}.$$

Thus, we can define

$$\boxed{\kappa_{ab} \triangleq -\frac{1}{2}\phi_{\mathrm{F}} \frac{m_{0z}}{M_S}}$$

where the Faraday rotation angle per unit length is

$$\phi_{\mathrm{F}} = \frac{1}{2}\frac{M_S f}{\sqrt{\varepsilon_r}}\omega\sqrt{\mu_0\varepsilon_0} = \frac{1}{2}\frac{M_S f k_0}{\sqrt{\varepsilon_r}}.$$

Problem 8.14 Since the coupling is proportional to the z component of the small-signal magnetization, it is useful to determine the relationship between the power per unit width carried by a forward volume spin wave and m_{0z}. Consider the z component of the small-signal magnetization in the limit $k_{\mathrm{SW}}d \ll 1$.

(a) Starting with the dispersion relation for forward volume waves (5.44),

$$\tan\left[\frac{k_{\mathrm{SW}}d}{2}\sqrt{-(1+\chi)} - \frac{n\pi}{2}\right] = \frac{1}{\sqrt{-(1+\chi)}}, \tag{8.88}$$

take $n = 0$ and show that

$$k_{\mathrm{SW}}d \approx \frac{2}{-(1+\chi)}. \tag{8.89}$$

(b) Show that $|\chi| \gg 1$.
(c) Using the results of parts (a) and (b), show that the normalization constant (6.33)

$$\psi_{0n} = \sqrt{\frac{4P_n}{-(1+\chi)\omega\mu_0 k_n d}} \tag{8.90}$$

reduces to

$$\psi_0 \approx \sqrt{\frac{2P}{\omega\mu_0}}, \tag{8.91}$$

and the small-signal magnetization

$$m_{0z} = \chi h_z = -\chi \partial_z \psi_0 = -ik_{SW} \chi \psi_0 \tag{8.92}$$

is approximated by

$$m_{0z} = i\frac{2}{d}\sqrt{\frac{2P}{\omega\mu_0}}. \tag{8.93}$$

Solution 8.14

(a) From (8.88), with $n = 0$,

$$\tan\left(\frac{k_{SW}d}{2}\sqrt{-(1+\chi)}\right) = \frac{1}{\sqrt{-(1+\chi)}}.$$

When we have $k_{SW}\, d\sqrt{-(1+\chi)} \ll 1$, the tangent function can be replaced with its argument,[3] leading to

$$\boxed{k_{SW}d \approx \frac{2}{-(1+\chi)}.}$$

(b) We also recognize that when $k_{SW}\, d \ll 1$,

$$1+\chi \gg 1 \text{ or } \boxed{|\chi| \gg 1.}$$

To further confirm this, note that $\omega \to \omega_0$ when $k_{SW}d \ll 1$. From (3.63a), we recall that χ has a singularity at $\omega = \omega_0$.

(c) The forward volume wave mode normalization coefficient is given by (6.33)

$$\psi_{0n} = \sqrt{\frac{4P_n}{-(1+\chi)\,\omega\mu_0 k_n d}}.$$

Substituting the result from (a) back into the expression for $\psi_{00} = \psi_0$ (with $k_n = k_{SW}$)

$$\boxed{\psi_0 \simeq \sqrt{\frac{2P}{\omega\mu_0}}.}$$

---

[3] Here we implicitly assume that in the limit $kd \to 0$, $kd\sqrt{-(1+\chi)} = 0$.

Problem 8.15 Calculate the conversion length

$$L_c = \frac{\pi M_S d}{|\phi_F|} \sqrt{\frac{\omega \mu_0}{8P}} \tag{8.94}$$

if $M_S = 140\,\text{kA/m}$, $d = 5\,\mu\text{m}$, $\omega/(2\pi) = 6\,\text{GHz}$, $P = 30\,\text{mW/mm}$, and $|\phi_F| = 0.17°/\mu\text{m}$. The conversion length is the distance over which one optical mode is completely converted into another mode.

Solution 8.15

$$
\begin{aligned}
L_c &= \frac{\pi M_S d}{|\phi_F|} \sqrt{\frac{\omega \mu_0}{8P}} \\
&= \frac{\pi \times 140 \times 10^3 \times 5 \times 10^{-6}}{0.17 \times 10^6 \times \pi/180} \times \sqrt{\frac{6 \times 10^9 \times 2\pi \times 4\pi \times 10^{-7}}{8 \times 30}}
\end{aligned}
$$

$$\therefore \quad \boxed{L_c = 10.4\,\text{mm}}$$

References

1. A. Yariv, P. Yeh, *Optical Waves in Crystals* (John Wiley, New York, 1984)
2. A. Yariv, *Quantum Electronics* (John Wiley, New York, 1989)
3. H.A. Haus, *Waves and Fields in Optoelectronics* (Prentice-Hall, Englewood Cliffs, 1984)
4. D.D. Stancil, A. Prabhakar, *Spin Waves: Theory and Applications* (Springer, New York, 2009)
5. H.A. Haus, Coupled mode theory. Proc. IEEE **79**(10), 1505 (1991)

Nonlinear Interactions

9

The theory of nonlinear spin wave interactions has a rich history that can be traced back to early work by Suhl [1]. As we increase the amplitude of the spin waves, the cone angle of precession becomes larger, and we must add nonlinear corrections to the linear small-signal form of the Landau–Lifshitz equation. One approach to general nonlinear problems in spin wave dynamics is to transform the Landau–Lifshitz torque equation of motion into a scalar Hamiltonian in terms of canonical variables. This was proposed by Schlömann [2], following which a generalized quantum mechanical description of nonlinear magnons (quantized spinwaves), was developed as the S-theory [4]. This helped expand our understanding of nonlinear spin wave excitations and led us to the nonlinear Schrödinger equation (NLSE). As we increase the spin wave amplitudes, the continuous wave signal breaks into an envelope of pulses. Under the right balance of nonlinearity and dispersion, we observe the formation of shape preserving pulses, or spin wave solitons.

This chapter begins with the Hamiltonian formalism for spin wave interactions in Problems 9.1–9.3. We then expand the Hamiltonian to include higher order nonlinear terms, and diagonalize the Hamiltonian, using the Bogoliubov transformation, in Problem 9.4. A semi-classical expansion, in Problem 9.5, allows us to derive the various coefficients in the nonlinear Hamiltonian. Problems 9.6–9.8 explore the conditions necessary for soliton formation and show that they are valid solutions to the NLSE .

9.1   Complex Spin Wave Amplitudes

We introduced the raising and lowering operators in Sect. 2.3. Here, we follow a similar derivation, beginning with the Hamiltonian equations of motion

$$\dot{q} = \frac{\partial \mathcal{H}}{\partial p}, \quad \dot{p} = -\frac{\partial \mathcal{H}}{\partial q}, \tag{9.1}$$

© The Author(s), under exclusive license to Springer Nature Switzerland AG 2021
D. D. Stancil and A. Prabhakar, *Spin Waves*,
https://doi.org/10.1007/978-3-030-68582-9_9

where the dot specifies the total time derivative and the Hamiltonian function is of the form

$$\mathcal{H} = \frac{p^2}{2m} + \frac{1}{2}kq^2. \tag{9.2}$$

The new canonical variables

$$a = \frac{1}{\sqrt{2}}(p + iq), \tag{9.3a}$$

$$a^* = \frac{1}{\sqrt{2}}(p - iq), \tag{9.3b}$$

represent the complex amplitudes of propagating spin waves. For small precession angles, $M_z \approx M_S$, and we use the resonant and anti-resonant eigenmodes of precession

$$M_\pm = M_x \pm i M_y, \tag{9.4a}$$

such that

$$M_x = \frac{1}{2}(M_+ + M_-), \tag{9.4b}$$

$$M_y = \frac{-i}{2}(M_+ - M_-), \tag{9.4c}$$

to find

$$a = M_+ [-2\gamma M_S]^{-1/2}, \tag{9.5a}$$
$$a^* = M_- [-2\gamma M_S]^{-1/2}. \tag{9.5b}$$

For larger precession angles, the transformation from $M_\pm$ to a, a^* will incorporate a scaling function that depends on the square of the amplitude $|a|^2 = a^* a$. Hence, we find a transformation

$$M_+ = af(a^*a)\sqrt{-2\gamma M_S}, \tag{9.6a}$$
$$M_- = a^* f(a^*a)\sqrt{-2\gamma M_S}, \tag{9.6b}$$

such that

$$\dot{a} = i\frac{\delta W}{\delta a^*}, \quad \dot{a}^* = -i\frac{\delta W}{\delta a}, \tag{9.7}$$

where W is the energy density

$$f = \sqrt{1 + \frac{\gamma aa^*}{2M_S}}, \tag{9.8}$$

and

$$M_z = M_S + \gamma a^* a. \tag{9.9}$$

The variational derivatives $\delta W/\delta a$, $\delta W/\delta a^*$ can be derived from the effective field

$$\mathbf{H}_{\text{eff}} = -\frac{1}{\mu_0}\frac{\delta W}{\delta \mathbf{M}} = -\frac{1}{\mu_0}\left[\frac{\delta W}{\delta M_x}\hat{\mathbf{x}} + \frac{\delta W}{\delta M_y}\hat{\mathbf{y}} + \frac{\delta W}{\delta M_z}\hat{\mathbf{z}}\right]. \tag{9.10}$$

and used in the lossless Landau–Lifshitz equation (3.51)

$$\dot{\mathbf{M}} = \gamma\mu_0\mathbf{M} \times \mathbf{H}_{\text{eff}}. \tag{9.11}$$

Problem 9.1 Show that the complex amplitude

$$a = \frac{1}{\sqrt{2}}(p + iq) \tag{9.12}$$

satisfies the equation of motion

$$\dot{a} = i\frac{\partial \mathcal{H}(a, a^*)}{\partial a^*}, \tag{9.13}$$

where p and q are canonical variables satisfying the Hamiltonian equations of motion in (9.1).

Solution 9.1
To obtain the equation of motion

$$\dot{a} = i\frac{\partial \mathcal{H}(a, a^*)}{\partial a^*},$$

we begin with the chain rule

$$\frac{\partial \mathcal{H}}{\partial a^*} = \frac{\partial \mathcal{H}}{\partial p}\frac{\partial p}{\partial a^*} + \frac{\partial \mathcal{H}}{\partial q}\frac{\partial q}{\partial a^*},$$

where $a = \frac{1}{\sqrt{2}}(p + iq)$ and $a^* = \frac{1}{\sqrt{2}}(p - iq)$. Adding and subtracting a and a^* allows us to obtain

$$p = \frac{1}{\sqrt{2}}\left(a + a^*\right),$$

$$q = \frac{i}{\sqrt{2}}\left(a^* - a\right).$$

Thus

$$\frac{\partial p}{\partial a^*} = \frac{1}{\sqrt{2}}, \quad \text{and} \quad \frac{\partial q}{\partial a^*} = \frac{i}{\sqrt{2}}.$$

From (9.1), we know

$$\dot{q} = \frac{\partial \mathcal{H}}{\partial p}, \quad \text{and} \quad \dot{p} = -\frac{\partial \mathcal{H}}{\partial q}.$$

The chain rule can now be written

$$\frac{\partial \mathcal{H}}{\partial a^*} = \dot{q} \frac{1}{\sqrt{2}} - \dot{p} \frac{i}{\sqrt{2}}$$

Multiplying both sides by i and collecting terms gives

$$\frac{1}{\sqrt{2}} (\dot{p} + i\dot{q}) = i \frac{\partial \mathcal{H}}{\partial a^*}$$

or,

$$\boxed{\dot{a} = i \frac{\partial \mathcal{H}}{\partial a^*}.}$$

Problem 9.2 In this problem, we will solve for the scaling function f in (9.6).

(a) Take the time derivative of (9.6a) and show that

$$\dot{M}_+ = \dot{a} f C + a f'[a^* \dot{a} + \dot{a}^* a] C \tag{9.14}$$

where $C = \sqrt{-2\gamma M_S}$ and $f'(x) = df/dx$.

(b) Using the canonical equations (9.7)

$$\dot{a} = i \frac{\delta W}{\delta a^*}, \quad \dot{a}^* = -i \frac{\delta W}{\delta a}, \tag{9.15}$$

for the time derivatives on the right-hand-side of (9.14), show that

$$\dot{M}_+ = iC \left[f + f' a^* a \right] \frac{\delta W}{\delta a^*} - iCa^2 f' \frac{\delta W}{\delta a}. \tag{9.16}$$

(c) Using the chain rules

$$\frac{\delta W}{\delta a} = \frac{\delta W}{\delta M_+} \frac{\partial M_+}{\partial a} + \frac{\delta W}{\delta M_-} \frac{\partial M_-}{\partial a}, \tag{9.17a}$$

$$\frac{\delta W}{\delta a^*} = \frac{\delta W}{\delta M_+} \frac{\partial M_+}{\partial a^*} + \frac{\delta W}{\delta M_-} \frac{\partial M_-}{\partial a^*}, \tag{9.17b}$$

show that

$$\frac{\delta W}{\delta a} = C(f + a^* a f') \frac{\delta W}{\delta M_+} + C(a^*)^2 f' \frac{\delta W}{\delta M_-}, \tag{9.18a}$$

$$\frac{\delta W}{\delta a^*} = Ca^2 f' \frac{\delta W}{\delta M_+} + C(f + a^* a f') \frac{\delta W}{\delta M_-}. \tag{9.18b}$$

(d) Substitute the results of part (c) into the result of part (b) and show that

$$\dot{M}_+ = -i2\gamma M_S \left(f^2 + 2ff'a^*a \right) \frac{\delta W}{\delta M_-}. \tag{9.19}$$

(e) Show that the result of part (d) has the same form as the equation of motion (9.13) provided f satisfies the differential equation

$$\frac{d}{dx}(xf^2) = f^2 + 2xff' = \sqrt{1 + \frac{2\gamma x f^2}{M_S}}, \tag{9.20}$$

where $x = a^*a$.

(f) Using direct substitution, show that

$$f(x) = \sqrt{1 + \frac{\gamma x}{2M_S}} \tag{9.21}$$

is a solution to the differential equation in part (e).

Solution 9.2

(a) We start with (9.6)

$$M_+ = af \left(a^*a \right) \sqrt{-2\gamma M_S} = aCf(a^*a) \tag{9.22}$$

where $C = \sqrt{-2\gamma M_S}$. Using the product and chain rules we know

$$\frac{dM_+}{dt} = \frac{da}{dt} fC + aC \frac{df(a^*a)}{d(a^*a)} \frac{d(a^*a)}{dt}$$

$$\boxed{\dot{M}_+ = \dot{a}fC + aCf' \left[\dot{a}^*a + a^*\dot{a} \right]}$$

(b) We are given that

$$\dot{a} = i \frac{\delta W}{\delta a^*}, \quad \dot{a}^* = -i \frac{\delta W}{\delta a}.$$

Substituting into the result from part (a) and collecting terms gives

$$\dot{M}_+ = fCi \frac{\delta W}{\delta a^*} + aCf' \left[-i \frac{\delta W}{\delta a} a + a^* i \frac{\delta W}{\delta a^*} \right]$$

$$\boxed{\dot{M}_+ = iC \left[f + f'a^*a \right] \frac{\delta W}{\delta a^*} - iCa^2 f' \frac{\delta W}{\delta a}}$$

(c) Let us look at each term in the chain rule

$$\frac{\delta W}{\delta a} = \frac{\delta W}{\delta M_+}\frac{\partial M_+}{\partial a} + \frac{\delta W}{\delta M_-}\frac{\partial M_-}{\partial a}.$$

From (9.6a)

$$\frac{\partial M_+}{\partial a} = fC + aCf'a^* = C\left(f + a^*af'\right).$$

Similarly, from (9.6b)

$$M_- = a^* f\left(a^* a\right)\sqrt{-2\gamma M_S}$$

$$\frac{\partial M_-}{\partial a} = a^*Cf'a^* = C\left(a^*\right)^2 f'.$$

The variational derivative then becomes

$$\boxed{\frac{\delta W}{\delta a} = \frac{\delta W}{\delta M_+}C\left(f + a^*af'\right) + \frac{\delta W}{\delta M_-}C\left(a^*\right)^2 f'.}$$

Similarly

$$\frac{\partial M_+}{\partial a^*} = aCf'a = Cf'a^2,$$

$$\frac{\partial M_-}{\partial a^*} = fC + a^*Cf'a = C\left(f + f'a^*a\right),$$

and

$$\frac{\delta W}{\delta a^*} = \frac{\delta W}{\delta M_+}\frac{\partial M_+}{\partial a^*} + \frac{\delta W}{\delta M_-}\frac{\partial M_-}{\partial a^*}$$

Thus

$$\boxed{\frac{\delta W}{\delta a^*} = \frac{\delta W}{\delta M_+}Cf'a^2 + \frac{\delta W}{\delta M_-}C\left(f + f'a^*a\right).}$$

(d) From (b)

$$\dot{M}_+ = iC\left(f + f'a^*a\right)\frac{\delta W}{\delta a^*} - iCa^2 f'\frac{\delta W}{\delta a}.$$

Substituting the results of part (c) gives

$$\dot{M}_+ = iC\left(f + f'a^*a\right)\left[\frac{\delta W}{\delta M_+}Cf'a^2 + \frac{\delta W}{\delta M_-}C\left(f + f'a^*a\right)\right]$$
$$- iCa^2 f'\left[\frac{\delta W}{\delta M_+}C\left(f + f'a^*a\right) + \frac{\delta W}{\delta M_-}C\left(a^*\right)^2 f'\right].$$

Simplifying again yields

$$
\begin{aligned}
\dot{M}_+ &= iC^2 \left[\frac{\delta W}{\delta M_+} \left(f'a^2 \left(f + f'a^*a \right) - f'a^2 \left(f + f'a^*a \right) \right) \right] \\
&\quad + iC^2 \left[\frac{\delta W}{\delta M_-} \left(\left(f + f'a^*a \right)^2 - a^2 (a^*)^2 \left(f' \right)^2 \right) \right] \\
&= iC^2 \frac{\delta W}{\delta M_-} \left[f^2 + 2ff'a^*a + \left(f'a^*a \right)^2 - \left(f'a^*a \right)^2 \right] \\
&= iC^2 \left(f^2 + 2ff'a^*a \right) \frac{\delta W}{\delta M_-}.
\end{aligned}
$$

Substituting for C^2, we get

$$
\boxed{ \dot{M}_+ = -i2\gamma M_S \left(f^2 + 2ff'a^*a \right) \frac{\delta W}{\delta M_-}. }
$$

(e) A comparison of the result of part (d) with the equation of motion (9.13)

$$
\dot{M}_+ = -2i\gamma M_z \frac{\delta W}{\delta M_-}
$$

shows that they will be of the same form if

$$
M_z = M_S \left(f^2 + 2ff'a^*a \right). \tag{9.23}
$$

But M_z can be written $M_z = \sqrt{M_S^2 - M_x^2 - M_y^2}$, or

$$
\frac{M_z}{M_S} = \sqrt{1 - \frac{\left(M_x^2 + M_y^2 \right)}{M_S^2}}.
$$

Note that

$$
M_+ M_- = \left(M_x + i M_y \right) \left(M_x - i M_y \right) = M_x^2 + M_y^2
$$

$$
\implies \frac{M_z}{M_S} = \sqrt{1 - \frac{M_+ M_-}{M_S^2}}.
$$

But

$$
\frac{M_+ M_-}{M_S^2} = \frac{aCfa^*Cf}{M_S^2} = \frac{(-2\gamma M_S) f^2 a^*a}{M_S^2} = \frac{-2\gamma f^2 a^*a}{M_S}
$$

and

$$
M_z = M_S \sqrt{1 + \frac{2\gamma f^2 a^*a}{M_S}}.
$$

Combining this with the previous expression for M_z (9.23) gives

$$f^2 + 2ff'a^*a = \sqrt{1 + \frac{2\gamma f^2 a^* a}{M_S}}.$$

Let $x = a^*a$, and recognize that

$$\frac{d}{dx}\left(xf^2\right) = f^2 + x2ff'.$$

We conclude that the result of part (d) is the same form as the equation of motion (9.13) provided f satisfies the differential equation

$$\boxed{\frac{d}{dx}\left(xf^2\right) = f^2 + 2xff' = \sqrt{1 + \frac{2\gamma f^2 x}{M_S}}.}$$

(f) Consider the trial solution (9.21)

$$f = \sqrt{1 + \frac{\gamma x}{2M_S}}.$$

Taking the derivative gives

$$f' = \frac{1}{2}\left(1 + \frac{\gamma x}{2M_S}\right)^{-1/2}\frac{\gamma}{2M_S} = \frac{\gamma}{4M_S}\frac{1}{f}.$$

Next, substitute into the differential equation from part (e)

$$1 + \frac{\gamma x}{2M_S} + 2xf\frac{\gamma}{4M_S}\frac{1}{f} \stackrel{?}{=} \sqrt{1 + \frac{2\gamma x}{M_S}\left(1 + \frac{\gamma x}{2M_S}\right)}$$

$$1 + \frac{\gamma x}{2M_S} + \frac{\gamma x}{2M_S} \stackrel{?}{=} \sqrt{1 + \frac{2\gamma x}{M_S} + \left(\frac{\gamma x}{M_S}\right)^2}$$

$$1 + \frac{\gamma x}{M_S} \stackrel{?}{=} \sqrt{\left(1 + \frac{\gamma x}{M_S}\right)^2}$$

Thus

$$\boxed{1 + \frac{\gamma x}{M_S} \stackrel{\checkmark}{=} \left(1 + \frac{\gamma x}{M_S}\right)}$$

We conclude that (9.21) is a solution to the differential equation found in part (e).

Problem 9.3 The magnetic energy density for small signal excitations about equilibrium ($\mathbf{M} = M_S \, \hat{\mathbf{z}}$) can be expanded into a quadratic variation

$$W = \mu_0 \frac{H_x M_x^2 + H_y M_y^2}{2M_S} \tag{9.24}$$

where H_x and H_y are the effective fields $\left(\dfrac{-1}{\mu_0} \dfrac{\partial W}{\partial \mathbf{M}}\right)$ in the $\hat{\mathbf{x}}$ and $\hat{\mathbf{y}}$ directions, respectively. Assuming a very thin film along $\hat{\mathbf{x}}$ with both applied field and uniaxial anisotropy along $\hat{\mathbf{z}}$, show that substituting the variables

$$M_x = \sqrt{\frac{|\gamma| M_S}{2}}(a^* + a), \quad M_y = i\sqrt{\frac{|\gamma| M_S}{2}}(a^* - a), \tag{9.25}$$

into W yields an expression that can be interpreted as a Hamiltonian of the form

$$\mathcal{H} = Aa^*a + \frac{B}{2}(aa + a^*a^*) \tag{9.26}$$

where $A = \mu_0 |\gamma| (H_x + H_y)/2$ and $B = \mu_0 |\gamma| (H_x - H_y)/2$.

Solution 9.3
Consider the energy density

$$W = \mu_0 \frac{H_x M_x^2 + H_y M_y^2}{2M_S}.$$

We are given that $M_x = \sqrt{\frac{|\gamma| M_S}{2}}(a^* + a)$, and $M_y = i\sqrt{\frac{|\gamma| M_S}{2}}(a^* - a)$. Substituting into W gives

$$
\begin{aligned}
W &= \mu_0 \frac{|\gamma| M_S}{2} \frac{(a^* + a)^2}{2M_S} H_x - \mu_0 \frac{|\gamma| M_S}{2} \frac{(a^* - a)^2}{2M_S} H_y \\
&= \frac{\mu_0 |\gamma|}{4} \left[\left[(a^*)^2 + 2a^*a + a^2 \right] H_x - \left[(a^*)^2 - 2a^*a + a^2 \right] H_y \right] \\
&= \frac{\mu_0 |\gamma|}{4} \left[2a^*a \left(H_x + H_y \right) + \left(a^*a^* + aa \right) \left(H_x - H_y \right) \right] \\
&= \underbrace{\frac{\mu_0 |\gamma|}{2} \left(H_x + H_y \right) a^*a}_{A} + \underbrace{\frac{\mu_0 |\gamma|}{4} \left(H_x - H_y \right)}_{B} \left(a^*a^* + aa \right).
\end{aligned}
$$

This can be interpreted as a Hamiltonian of the form

$$\boxed{\mathcal{H} = Aa^*a + \frac{B}{2} \left(a^*a^* + aa \right).}$$

9.2   Bogoliubov Transformation

The energy density W when expressed in terms of a and a^* becomes the Hamiltonian function $\mathcal{H}(a, a^*)$. Let us expand $\mathcal{H}(a, a^*)$ in a power series, with each term in the series being a function of the canonical variables a and a^*

$$\mathcal{H}(a, a^*) = \mathcal{H}^{(0)} + \mathcal{H}^{(1)} + \mathcal{H}^{(2)} + \ldots \tag{9.27}$$

In the absence of spin waves, the medium is in equilibrium and $\mathcal{H}(a^*, a)$ has a minimum for $a = a^* = 0$ such that the first order term $\mathcal{H}^{(1)} = 0$. Hence, we begin with contributions to the Hamiltonian that are second order in a and a^*, written as[1]

$$\mathcal{H}^{(2)} = \sum_{\mathbf{k}} \left(A_{\mathbf{k}} a_{\mathbf{k}}^* a_{\mathbf{k}} + \frac{1}{2} [B_{\mathbf{k}} a_{\mathbf{k}}^* a_{-\mathbf{k}}^* + B_{\mathbf{k}}^* a_{\mathbf{k}} a_{-\mathbf{k}}] \right) \tag{9.28}$$

where $a_{\mathbf{k}}$ is the spatial Fourier transform of a

$$a_{\mathbf{k}}(t) = \frac{1}{V} \int a(\mathbf{r}, t) e^{-i\mathbf{k}\cdot\mathbf{r}} d^3r. \tag{9.29}$$

We will determine the coefficients $A_{\mathbf{k}}$ and $B_{\mathbf{k}}$ in Problem 9.5. The Hamiltonian equations of motion (also referred to as the "rate equations") now become

$$\dot{a}_{\mathbf{k}} = i \frac{\partial \mathcal{H}}{\partial a_{\mathbf{k}}^*}, \quad \dot{a}_{\mathbf{k}}^* = -i \frac{\partial \mathcal{H}}{\partial a_{\mathbf{k}}}. \tag{9.30}$$

Considering only the first term in $\mathcal{H}^{(2)}$, the equations in (9.30) are written as

$$\dot{a}_{\mathbf{k}} = i A_{\mathbf{k}} a_{\mathbf{k}}, \quad \dot{a}_{\mathbf{k}}^* = -i A_{\mathbf{k}} a_{\mathbf{k}}^*, \tag{9.31}$$

which have time harmonic solutions of the form $e^{\pm i\omega_{\mathbf{k}} t}$, where $\omega_{\mathbf{k}} \equiv A_{\mathbf{k}}$. However, if we include the second term of (9.28) and evaluate (9.30), the variables $a_{\mathbf{k}}$ and $a_{\mathbf{k}}^*$ will represent elliptical precession, not circular. To recover the resonant circular form, we invoke a second change of variables

$$b_{\mathbf{k}} = u_{\mathbf{k}} a_{\mathbf{k}} + v_{\mathbf{k}} a_{-\mathbf{k}}^*, \tag{9.32a}$$

$$b_{\mathbf{k}}^* = u_{\mathbf{k}}^* a_{\mathbf{k}}^* + v_{\mathbf{k}}^* a_{-\mathbf{k}}, \tag{9.32b}$$

where[2]

$$|u_{\mathbf{k}}|^2 - |v_{\mathbf{k}}|^2 = 1. \tag{9.33}$$

---

[1] Here $a_{\mathbf{k}}^* \equiv (a_{\mathbf{k}})^* = (a^*)_{-\mathbf{k}}$.
[2] We will see in Problem 9.4 that $u_{\mathbf{k}}$ can be chosen to be real without loss of generality.

This is the *Bogoliubov transformation*, and the transformation coefficients u_k and v_k are chosen so that the Hamiltonian is transformed to a diagonal form

$$\mathcal{H}^{(2)} = \sum_k \omega_k b_k^* b_k. \tag{9.34}$$

It is worth pausing to note that the higher order terms $\mathcal{H}^{(3)}$ and $\mathcal{H}^{(4)}$ in the Hamiltonian give rise to many nonlinear effects such as modulation instability, frequency doubling and auto-oscillations [6]. The onset of such phenomena is typically determined by the threshold for the first and second Suhl instability, in the case of 3 and 4 magnon processes, respectively. These processes are also understood using the canonical transformations that diagonalize (9.27) [7].

Problem 9.4 Prove the results of the Bogoliubov transformation

(a) Substituting (9.32)

$$b_k = u_k a_k + v_k a_{-k}^* \tag{9.35a}$$
$$b_k^* = u_k^* a_k^* + v_k^* a_{-k} \tag{9.35b}$$

into (9.34)

$$\mathcal{H}^{(2)} = \sum_k \omega_k b_k^* b_k, \tag{9.36}$$

show that we obtain the general (non-diagonal) form of the second order Hamiltonian given in (9.28):

$$\mathcal{H}^{(2)} = \sum_k \left(A_k a_k^* a_k + \frac{1}{2} [B_k a_k^* a_{-k}^* + B_k^* a_k a_{-k}] \right). \tag{9.37}$$

Identify the appropriate terms to yield

$$\omega_k \left[u_k^2 + |v_k|^2 \right] = A_k, \tag{9.38a}$$
$$2\omega_k u_k v_k = B_k, \tag{9.38b}$$
$$2\omega_k u_k v_k^* = B_k^*. \tag{9.38c}$$

(Hint: you may use the properties $u_k = u_{-k}$, $v_k = v_{-k}$.)

(b) Using the condition

$$|u_k|^2 - |v_k|^2 = 1 \tag{9.39}$$

while solving (9.38), show that

$$u_k = \sqrt{\frac{A_k + \omega_k}{2\omega_k}}, \quad \text{and} \quad v_k = \frac{B_k}{|B_k|}\sqrt{\frac{A_k - \omega_k}{2\omega_k}}. \tag{9.40}$$

(c) Use (9.36) to write down, in matrix notation, the rate equations for $b_\mathbf{k}$ and $b^*_{-\mathbf{k}}$. Show that the transformation

$$\begin{pmatrix} b_\mathbf{k} \\ b^*_{-\mathbf{k}} \end{pmatrix} = \begin{pmatrix} u_\mathbf{k} & v_\mathbf{k} \\ v^*_\mathbf{k} & u_\mathbf{k} \end{pmatrix} \begin{pmatrix} a_\mathbf{k} \\ a^*_{-\mathbf{k}} \end{pmatrix}, \tag{9.41}$$

will yield the corresponding rate equations $\left(\dot{a}_\mathbf{k} \ \dot{a}^*_{-\mathbf{k}} \right)^T$

$$\begin{bmatrix} \dot{a}_\mathbf{k} \\ \dot{a}^*_{-\mathbf{k}} \end{bmatrix} = i \begin{bmatrix} A_\mathbf{k} & B_\mathbf{k} \\ -B^*_\mathbf{k} & -A_\mathbf{k} \end{bmatrix} \begin{bmatrix} a_\mathbf{k} \\ a^*_{-\mathbf{k}} \end{bmatrix}. \tag{9.42}$$

(Hint: You may use the fact that $\omega_\mathbf{k} = \omega_{-\mathbf{k}}$.)

Solution 9.4

(a) Start with the Hamiltonian

$$\mathcal{H}^{(2)} = \sum_\mathbf{k} \omega_\mathbf{k} b^*_\mathbf{k} b_\mathbf{k}$$

where

$$b_\mathbf{k} = u_\mathbf{k} a_\mathbf{k} + v_\mathbf{k} a^*_{-\mathbf{k}},$$
$$b^*_\mathbf{k} = u^*_\mathbf{k} a^*_\mathbf{k} + v^*_\mathbf{k} a_{-\mathbf{k}}.$$

Substituting gives

$$\mathcal{H}^{(2)} = \sum_\mathbf{k} \omega_\mathbf{k} \left(u^*_\mathbf{k} a^*_\mathbf{k} + v^*_\mathbf{k} a_{-\mathbf{k}} \right) \left(u_\mathbf{k} a_\mathbf{k} + v_\mathbf{k} a^*_{-\mathbf{k}} \right)$$
$$= \sum_\mathbf{k} \omega_\mathbf{k} \left[|u_\mathbf{k}|^2 a^*_\mathbf{k} a_\mathbf{k} + u_\mathbf{k} v^*_\mathbf{k} a_{-\mathbf{k}} a_\mathbf{k} + u^*_\mathbf{k} v_\mathbf{k} a^*_\mathbf{k} a^*_{-\mathbf{k}} + |v_\mathbf{k}|^2 a^*_{-\mathbf{k}} a_{-\mathbf{k}} \right]$$

Since the sum is over all values of $\mathbf{k}$, the sum over $|v_\mathbf{k}|^2 a^*_{-\mathbf{k}} a_{-\mathbf{k}}$ will give the same result as the sum over $|v_{-\mathbf{k}}|^2 a^*_\mathbf{k} a_\mathbf{k}$. Thus, making the substitution $\mathbf{k} \to -\mathbf{k}$ in the last term allows us to write

$$\mathcal{H}^{(2)} = \sum_\mathbf{k} \omega_\mathbf{k} \left[\left(|u_\mathbf{k}|^2 + |v_{-\mathbf{k}}|^2 \right) a^*_\mathbf{k} a_\mathbf{k} + u^*_\mathbf{k} v_\mathbf{k} a^*_\mathbf{k} a^*_{-\mathbf{k}} + u_\mathbf{k} v^*_\mathbf{k} a_{-\mathbf{k}} a_\mathbf{k} \right].$$

Suppose that we factored out a phase factor from $b_\mathbf{k} = u_\mathbf{k} a_\mathbf{k} + v_\mathbf{k} a^*_{-\mathbf{k}}$ such that $u_\mathbf{k}$ was real. Since the phase factor would cancel when we construct the Hamiltonian term $b^*_\mathbf{k} b_\mathbf{k}$, we conclude that we can take $u_\mathbf{k}$ to be real without loss of generality. Comparing with the form of (9.37)

$$\mathcal{H}^{(2)} = \sum_{\mathbf{k}} \left[A_{\mathbf{k}} a_{\mathbf{k}}^* a_{\mathbf{k}} + \frac{1}{2} \left(B_{\mathbf{k}} a_{\mathbf{k}}^* a_{-\mathbf{k}}^* + B_{\mathbf{k}}^* a_{\mathbf{k}} a_{-\mathbf{k}} \right) \right]$$

we obtain

$$\boxed{\begin{aligned} \omega_{\mathbf{k}} \left[u_{\mathbf{k}}^2 + |v_{-\mathbf{k}}|^2 \right] &= A_{\mathbf{k}} \\ 2\omega_{\mathbf{k}} u_{\mathbf{k}} v_{\mathbf{k}} &= B_{\mathbf{k}} \\ 2\omega_{\mathbf{k}} u_{\mathbf{k}} v_{\mathbf{k}}^* &= B_{\mathbf{k}}^* \end{aligned}} \qquad (9.43)$$

These equations are identical with (9.38) provided $v_{\mathbf{k}} = v_{-\mathbf{k}}$.

(b) We begin with equations (9.38)

$$\omega_{\mathbf{k}} \left[u_{\mathbf{k}}^2 + |v_{\mathbf{k}}|^2 \right] = A_{\mathbf{k}} \qquad (9.44)$$

$$2\omega_{\mathbf{k}} u_{\mathbf{k}} v_{\mathbf{k}} = B_{\mathbf{k}}$$

$$2\omega_{\mathbf{k}} u_{\mathbf{k}} v_{\mathbf{k}}^* = B_{\mathbf{k}}^*$$

along with the condition[3] from (9.39)

$$u_{\mathbf{k}}^2 - |v_{\mathbf{k}}|^2 = 1, \qquad (9.45)$$

where $u_{\mathbf{k}}$ is taken to be real. Substituting (9.45) into (9.44) gives

$$\left[u_{\mathbf{k}}^2 + u_{\mathbf{k}}^2 - 1 \right] = \frac{A_{\mathbf{k}}}{\omega_{\mathbf{k}}}$$

$$2u_{\mathbf{k}}^2 = \frac{A_{\mathbf{k}}}{\omega_{\mathbf{k}}} + 1$$

$$u_{\mathbf{k}}^2 = \frac{A_{\mathbf{k}} + \omega_{\mathbf{k}}}{2\omega_{\mathbf{k}}}$$

$$\boxed{u_{\mathbf{k}} = \sqrt{\frac{A_{\mathbf{k}} + \omega_{\mathbf{k}}}{2\omega_{\mathbf{k}}}}.}$$

Here we have chosen the positive square root, since the phase factor that was chosen to make $u_{\mathbf{k}}$ real can also be chosen to ensure $u_{\mathbf{k}} > 0$.

---

[3]This condition ensures that the transformation is *symplectic*.

Substituting (9.45) into (9.44) but solving instead for $|v_\mathbf{k}|$ gives

$$\left[1 + 2|v_\mathbf{k}|^2\right] = \frac{A_\mathbf{k}}{\omega_\mathbf{k}}$$

$$|v_\mathbf{k}| = \sqrt{\frac{A_\mathbf{k} - \omega_\mathbf{k}}{2\omega_\mathbf{k}}}.$$

To recover the phase of $v_\mathbf{k}$, we note from (9.38b) that

$$v_\mathbf{k} = \frac{B_\mathbf{k}}{2\omega_\mathbf{k} u_\mathbf{k}} = \frac{B_\mathbf{k}}{|B_\mathbf{k}|} \frac{|B_\mathbf{k}|}{2\omega_\mathbf{k} u_\mathbf{k}}$$

$$v_\mathbf{k} = \frac{B_\mathbf{k}}{|B_\mathbf{k}|} |v_\mathbf{k}|.$$

Note that $u_\mathbf{k} = |u_\mathbf{k}|$ since the phase factor has been chosen to be positive definite. Thus

$$\boxed{v_\mathbf{k} = \frac{B_\mathbf{k}}{|B_\mathbf{k}|} \sqrt{\frac{A_\mathbf{k} - \omega_\mathbf{k}}{2\omega_\mathbf{k}}}.}$$

(c) With $\mathcal{H} = \sum_{\mathbf{k}'} \omega_{\mathbf{k}'} b_{\mathbf{k}'}^* b_{\mathbf{k}'}$, Hamilton's equations become

$$\dot{b}_\mathbf{k} = i \frac{\partial}{\partial b_\mathbf{k}^*} \sum_{\mathbf{k}'} \omega_{\mathbf{k}'} b_{\mathbf{k}'}^* b_{\mathbf{k}'} = i \omega_\mathbf{k} b_\mathbf{k}$$

$$\dot{b}_\mathbf{k}^* = -i \frac{\partial}{\partial b_\mathbf{k}} \sum_{\mathbf{k}'} \omega_{\mathbf{k}'} b_{\mathbf{k}'}^* b_{\mathbf{k}'} = -i \omega_\mathbf{k} b_\mathbf{k}^*$$

This can be written in matrix notation as

$$\begin{bmatrix} \dot{b}_\mathbf{k} \\ \dot{b}_\mathbf{k}^* \end{bmatrix} = i\omega_\mathbf{k} \begin{bmatrix} 1 & 0 \\ 0 & -1 \end{bmatrix} \begin{bmatrix} b_\mathbf{k} \\ b_\mathbf{k}^* \end{bmatrix}.$$

$b_\mathbf{k}$ and $b_\mathbf{k}^*$ can be related to $a_\mathbf{k}$ and $a_\mathbf{k}^*$ using the Bogoliubov transformation

$$b_\mathbf{k} = u_\mathbf{k} a_\mathbf{k} + v_\mathbf{k} a_{-\mathbf{k}}^*$$
$$b_\mathbf{k}^* = u_\mathbf{k} a_\mathbf{k}^* + v_\mathbf{k}^* a_{-\mathbf{k}}$$

However, these cannot be written in matrix notation as they stand. To address this, we make the substitution $\mathbf{k} \to -\mathbf{k}$ in the second equation, and use the facts that $u_\mathbf{k} = u_{-\mathbf{k}}$ and $v_\mathbf{k} = v_{-\mathbf{k}}$. This gives

$$b_{\mathbf{k}} = u_{\mathbf{k}} a_{\mathbf{k}} + v_{\mathbf{k}} a^*_{-\mathbf{k}}$$
$$b^*_{-\mathbf{k}} = v^*_{\mathbf{k}} a_{\mathbf{k}} + u_{\mathbf{k}} a^*_{-\mathbf{k}}$$

or

$$\begin{bmatrix} b_{\mathbf{k}} \\ b^*_{-\mathbf{k}} \end{bmatrix} = \begin{bmatrix} u_{\mathbf{k}} & v_{\mathbf{k}} \\ v^*_{\mathbf{k}} & u_{\mathbf{k}} \end{bmatrix} \begin{bmatrix} a_{\mathbf{k}} \\ a^*_{-\mathbf{k}} \end{bmatrix}$$

Inverting this gives

$$\begin{bmatrix} a_{\mathbf{k}} \\ a^*_{-\mathbf{k}} \end{bmatrix} = \frac{1}{u_{\mathbf{k}}^2 - |v_{\mathbf{k}}|^2} \begin{bmatrix} u_{\mathbf{k}} & -v_{\mathbf{k}} \\ -v^*_{\mathbf{k}} & u_{\mathbf{k}} \end{bmatrix} \begin{bmatrix} b_{\mathbf{k}} \\ b^*_{-\mathbf{k}} \end{bmatrix} = \begin{bmatrix} u_{\mathbf{k}} & -v_{\mathbf{k}} \\ -v^*_{\mathbf{k}} & u_{\mathbf{k}} \end{bmatrix} \begin{bmatrix} b_{\mathbf{k}} \\ b^*_{-\mathbf{k}} \end{bmatrix}$$

where we have used the condition (9.45): $u_{\mathbf{k}}^2 - |v_{\mathbf{k}}|^2 = 1$. The rate equations (i.e., the equations of motion) for $a_{\mathbf{k}}, a^*_{-\mathbf{k}}$ are, therefore

$$\begin{bmatrix} \dot{a}_{\mathbf{k}} \\ \dot{a}^*_{-\mathbf{k}} \end{bmatrix} = \begin{bmatrix} u_{\mathbf{k}} & -v_{\mathbf{k}} \\ -v^*_{\mathbf{k}} & u_{\mathbf{k}} \end{bmatrix} \begin{bmatrix} \dot{b}_{\mathbf{k}} \\ \dot{b}^*_{-\mathbf{k}} \end{bmatrix}$$
$$= i\omega_{\mathbf{k}} \begin{bmatrix} u_{\mathbf{k}} & -v_{\mathbf{k}} \\ -v^*_{\mathbf{k}} & u_{\mathbf{k}} \end{bmatrix} \begin{bmatrix} 1 & 0 \\ 0 & -1 \end{bmatrix} \begin{bmatrix} b_{\mathbf{k}} \\ b^*_{-\mathbf{k}} \end{bmatrix}$$
$$= i\omega_{\mathbf{k}} \begin{bmatrix} u_{\mathbf{k}} & v_{\mathbf{k}} \\ -v^*_{\mathbf{k}} & -u_{\mathbf{k}} \end{bmatrix} \begin{bmatrix} u_{\mathbf{k}} & v_{\mathbf{k}} \\ v^*_{\mathbf{k}} & u_{\mathbf{k}} \end{bmatrix} \begin{bmatrix} a_{\mathbf{k}} \\ a^*_{-\mathbf{k}} \end{bmatrix}$$
$$= i\omega_{\mathbf{k}} \begin{bmatrix} u_{\mathbf{k}}^2 + |v_{\mathbf{k}}|^2 & 2u_{\mathbf{k}} v_{\mathbf{k}} \\ -2u_{\mathbf{k}} v^*_{\mathbf{k}} & -(u_{\mathbf{k}}^2 + |v_{\mathbf{k}}|^2) \end{bmatrix} \begin{bmatrix} a_{\mathbf{k}} \\ a^*_{-\mathbf{k}} \end{bmatrix}.$$

Note that in the second line we have used $\omega_{\mathbf{k}} = \omega_{-\mathbf{k}}$. This gives the desired result

$$\begin{bmatrix} \dot{a}_{\mathbf{k}} \\ \dot{a}^*_{-\mathbf{k}} \end{bmatrix} = i \begin{bmatrix} A_{\mathbf{k}} & B_{\mathbf{k}} \\ -B^*_{\mathbf{k}} & -A_{\mathbf{k}} \end{bmatrix} \begin{bmatrix} a_{\mathbf{k}} \\ a^*_{-\mathbf{k}} \end{bmatrix}.$$

Problem 9.5 The Landau–Lifshitz equation of motion can be used to derive the various coefficients in the Hamiltonian for the spin wave manifold. From (3.57), the linearized equation of motion can be written

$$\dot{\mathbf{m}} = \hat{\mathbf{z}} \times [-\omega_M \mathbf{h} + \omega_0 \mathbf{m}], \tag{9.46}$$

or

$$\dot{m}_x = \omega_M h_y - \omega_0 m_y, \tag{9.47a}$$
$$\dot{m}_y = -\omega_M h_x + \omega_0 m_x. \tag{9.47b}$$

For a uniform plane wave, the small-signal magnetic field intensity is approximately (see (5.21))

$$\mathbf{h_k} = -\frac{\mathbf{kk} \cdot \mathbf{m_k}}{k^2}.$$ (9.48)

Let us consider the direction of $\mathbf{k}$ to be specified by the spherical coordinate angles θ and ϕ so that

$$\hat{\mathbf{k}} = \hat{\mathbf{x}} \sin\theta\cos\phi + \hat{\mathbf{y}} \sin\theta\sin\phi + \hat{\mathbf{z}}\cos\theta.$$ (9.49)

Using (9.48), the transverse components of $\mathbf{h}$ needed for the linearized equation of motion are

$$h_{\mathbf{k}x} = -\left[\sin^2\theta\cos^2\phi\, m_{\mathbf{k}x} + \frac{1}{2}\sin^2\theta\sin(2\phi)\, m_{\mathbf{k}y} \right],$$ (9.50a)

$$h_{\mathbf{k}y} = -\left[\frac{1}{2}\sin^2\theta\sin(2\phi)\, m_{\mathbf{k}x} + \sin^2\theta\sin^2\phi\, m_{\mathbf{k}y} \right].$$ (9.50b)

Substituting these expressions for the magnetic field intensity into the linearized equation of motion (9.47) and collecting terms gives

$$\dot{m}_{\mathbf{k}x} = -\frac{\omega_M}{2}\sin^2\theta\sin(2\phi)m_{\mathbf{k}x} - (\omega_0 + \omega_M\sin^2\theta\sin^2\phi)m_{\mathbf{k}y},$$ (9.51a)

$$\dot{m}_{\mathbf{k}y} = (\omega_0 + \omega_M\sin^2\theta\cos^2\phi)m_{\mathbf{k}x} + \frac{\omega_M}{2}\sin^2\theta\sin(2\phi)m_{\mathbf{k}y}.$$ (9.51b)

Using

$$a_{\mathbf{k}}(t) = \frac{1}{M_S}\left(m_{\mathbf{k}x}(t) + i m_{\mathbf{k}y}(t)\right),$$ (9.52a)

$$a^*_{-\mathbf{k}}(t) = \frac{1}{M_S}\left(m_{\mathbf{k}x}(t) - i m_{\mathbf{k}y}(t)\right),$$ (9.52b)

write the small signal magnetization in terms of the canonical variables as

$$\frac{m_{\mathbf{k}x}}{M_S} = \frac{a_{\mathbf{k}} + a^*_{-\mathbf{k}}}{2} \quad \text{and} \quad \frac{m_{\mathbf{k}y}}{M_S} = \frac{a_{\mathbf{k}} - a^*_{-\mathbf{k}}}{2i}.$$ (9.53)

Substitute this into (9.51) to derive $A_{\mathbf{k}}$ and $B_{\mathbf{k}}$ as

$$A_{\mathbf{k}} = \omega_0 + \frac{1}{2}\omega_M\sin^2\theta,$$ (9.54a)

$$B_{\mathbf{k}} = \frac{1}{2}\omega_M\sin^2\theta e^{i2\phi}.$$ (9.54b)

Solution 9.5
From (9.52)

$$a_{\mathbf{k}}(t) = \frac{1}{M_S}\left(m_{\mathbf{k}x}(t) + i m_{\mathbf{k}y}(t)\right),$$

$$a^*_{-\mathbf{k}}(t) = \frac{1}{M_S}\left(m_{\mathbf{k}x}(t) - i m_{\mathbf{k}y}(t)\right).$$

Adding and subtracting the equations gives

$$\frac{m_{\mathbf{k}x}}{M_S} = \frac{a_{\mathbf{k}} + a^*_{-\mathbf{k}}}{2},$$

$$\frac{m_{\mathbf{k}y}}{M_S} = \frac{a_{\mathbf{k}} - a^*_{-\mathbf{k}}}{2i}.$$

From (9.51)

$$\dot{m}_{\mathbf{k}x} = -\frac{\omega_M}{2}\sin^2\theta \sin(2\phi)\, m_{\mathbf{k}x} - \left(\omega_0 + \omega_M \sin^2\phi \sin^2\theta\right) m_{\mathbf{k}y} \tag{9.55a}$$

$$\dot{m}_{\mathbf{k}y} = \left(\omega_0 + \omega_M \sin^2\theta \cos^2\phi\right) m_{\mathbf{k}x} + \frac{\omega_M}{2}\sin^2\theta \sin(2\phi)\, m_{\mathbf{k}y} \tag{9.55b}$$

Substituting for $\dfrac{m_{\mathbf{k}x}}{M_S}$ and $\dfrac{m_{\mathbf{k}y}}{M_S}$ into (9.55a) gives

$$\frac{\dot{m}_{\mathbf{k}x}}{M_S} = -\frac{\omega_M}{2}\sin^2\theta \sin(2\phi)\frac{m_{\mathbf{k}x}}{M_S} - \left(\omega_0 + \omega_M \sin^2\phi \sin^2\theta\right)\frac{m_{\mathbf{k}y}}{M_S},$$

or,

$$\frac{\dot{a}_{\mathbf{k}} + \dot{a}^*_{-\mathbf{k}}}{2} = \frac{-\omega_M}{2}\sin^2\theta \sin(2\phi)\left(\frac{a_{\mathbf{k}} + a^*_{-\mathbf{k}}}{2}\right) - \left(\omega_0 + \omega_M \sin^2\phi \sin^2\theta\right)\left(\frac{a_{\mathbf{k}} - a^*_{-\mathbf{k}}}{2i}\right).$$

Multiplying throughout by 2 and collecting terms gives

$$\dot{a}_{\mathbf{k}} + \dot{a}^*_{-\mathbf{k}} = \left[-\frac{\omega_M}{2}\sin^2\theta \sin(2\phi) + i\left(\omega_0 + \omega_M \sin^2\phi \sin^2\theta\right)\right]a_{\mathbf{k}}$$

$$+ \left[-\frac{\omega_M}{2}\sin^2\theta \sin(2\phi) - i\left(\omega_0 + \omega_M \sin^2\phi \sin^2\theta\right)\right]a^*_{-\mathbf{k}}$$

$$\tag{9.56}$$

Similarly, substituting for $\dfrac{m_{\mathbf{k}x}}{M_S}$ and $\dfrac{m_{\mathbf{k}y}}{M_S}$ into (9.55b) gives

$$\frac{\dot{a}_{\mathbf{k}} - \dot{a}^*_{-\mathbf{k}}}{2i} = \left(\omega_0 + \omega_M \sin^2\theta \cos^2\phi\right)\left(\frac{a_{\mathbf{k}} + a^*_{-\mathbf{k}}}{2}\right) + \frac{\omega_M}{2}\sin^2\theta \sin(2\phi)\left(\frac{a_{\mathbf{k}} - a^*_{-\mathbf{k}}}{2i}\right)$$

or,

$$\dot{a}_{\mathbf{k}} - \dot{a}^*_{-\mathbf{k}} = \left[i \left(\omega_0 + \omega_M \sin^2 \theta \cos^2 \phi \right) + \frac{\omega_M}{2} \sin^2 \theta \sin (2\phi) \right] a_{\mathbf{k}}$$

$$+ \left[i \left(\omega_0 + \omega_M \sin^2 \theta \cos^2 \phi \right) - \frac{\omega_M}{2} \sin^2 \theta \sin (2\phi) \right] a^*_{-\mathbf{k}}. \quad (9.57)$$

Adding (9.56) and (9.57) gives

$$2\dot{a}_{\mathbf{k}} = i \left[2\omega_0 + \omega_M \sin^2 \theta \left(\cos^2 \theta + \sin^2 \theta \right) \right] a_{\mathbf{k}}$$

$$+ \omega_M \sin^2 \theta \left[-\sin (2\phi) + i \left(\cos^2 \phi - \sin^2 \phi \right) \right] a^*_{-\mathbf{k}}.$$

But $\cos^2 \theta + \sin^2 \theta = 1$, and $\cos^2 \phi - \sin^2 \phi = \cos^2 (2\phi)$. Making these substitutions and dividing by 2 gives

$$\dot{a}_{\mathbf{k}} = i \left(\omega_0 + \frac{\omega_M}{2} \sin^2 \theta \right) a_{\mathbf{k}} + i \frac{\omega_M}{2} \sin^2 \theta \left(i \sin (2\phi) + \cos (2\phi) \right) a^*_{-\mathbf{k}}$$

$$= i \left(\omega_0 + \frac{\omega_M}{2} \sin^2 \theta \right) a_{\mathbf{k}} + i \frac{\omega_M}{2} \sin^2 \theta e^{i2\phi} a^*_{-\mathbf{k}}. \quad (9.58)$$

Subtracting (9.56) and (9.57) gives

$$2\dot{a}^*_{-\mathbf{k}} = \left[-\omega_M \sin^2 \theta \sin (2\phi) + i\omega_M \sin^2 \theta \left(\sin^2 \phi - \cos^2 \phi \right) \right] a_{\mathbf{k}}$$

$$- i \left[2\omega_0 + \omega_M \sin^2 \theta \left(\sin^2 \phi + \cos^2 \phi \right) \right] a^*_{-\mathbf{k}}$$

i.e.,

$$\dot{a}^*_{-\mathbf{k}} = -i \frac{\omega_M}{2} \sin^2 \theta \left(\cos (2\phi) - i \sin (2\phi) \right) a_{\mathbf{k}} - i \left(\omega_0 + \frac{\omega_M}{2} \sin^2 \theta \right) a^*_{-\mathbf{k}}$$

$$= -i \frac{\omega_M}{2} \sin^2 \theta e^{-i2\phi} a_{\mathbf{k}} - i \left(\omega_0 + \frac{\omega_M}{2} \sin^2 \theta \right) a^*_{-\mathbf{k}}. \quad (9.59)$$

Writing (9.58) and (9.59) in matrix form gives

$$\begin{bmatrix} \dot{a}_{\mathbf{k}} \\ \dot{a}^*_{-\mathbf{k}} \end{bmatrix} = i \begin{bmatrix} \omega_0 + \frac{\omega_M}{2} \sin^2 \theta & \frac{\omega_M}{2} \sin^2 \theta e^{i2\phi} \\ -\frac{\omega_M}{2} \sin^2 \theta e^{-i2\phi} & -\left(\omega_0 + \frac{\omega_M}{2} \sin^2 \theta \right) \end{bmatrix} \begin{bmatrix} a_{\mathbf{k}} \\ a^*_{-\mathbf{k}} \end{bmatrix}.$$

Comparing this to (9.42) from Problem 9.4

$$\frac{d}{dt} \begin{bmatrix} a_{\mathbf{k}} \\ a^*_{-\mathbf{k}} \end{bmatrix} = i \begin{bmatrix} A_{\mathbf{k}} & B_{\mathbf{k}} \\ -B^*_{\mathbf{k}} & -A_{\mathbf{k}} \end{bmatrix} \begin{bmatrix} a_{\mathbf{k}} \\ a^*_{-\mathbf{k}} \end{bmatrix}$$

we conclude

$$\boxed{\begin{array}{l} A_{\mathbf{k}} = \omega_0 + \frac{1}{2}\omega_M \sin^2 \theta \\[2mm] B_{\mathbf{k}} = \frac{1}{2} \sin^2 \omega_M \theta e^{i2\phi} \end{array}}$$

9.3   Nonlinear Schrödinger Equation

Beyond the onset of nonlinearities, the linear solutions to the Landau–Lifshitz equation are no longer valid. Instead, we use a general form for the dispersion relation

$$G(\omega, \mathbf{k}) = G\left(\omega(k, |\psi|^2), k, |\psi|^2\right) = 0, \tag{9.60}$$

where $|\psi|^2 = (1/2)(m/M_S)^2$, accounts for the reduction in magnetization along the direction of the applied bias field. A Taylor series expansion, up to second order corrections,

$$\begin{aligned}
\delta\omega &= \left(\frac{\partial\omega}{\partial k}\right)\delta k + \frac{1}{2}\frac{\partial^2\omega}{\partial k^2}\delta k^2 + \left(\frac{\partial\omega}{\partial|\psi|^2}\right)|\psi|^2 \\
&= v_g\delta k + \frac{D}{2}\delta k^2 + N|\psi|^2,
\end{aligned} \tag{9.61}$$

gives us the change in frequency, where we have defined the *dispersion* and *nonlinearity coefficients* D and N, respectively. Assuming propagation of the form $\psi(z, t) = \psi_0 \exp(i(\omega_0 t - k_0 z))$, we note that $\delta\omega \to i\partial/\partial t$ and $\delta k \to -i\partial/\partial z$, and equation (9.61) can be rewritten as

$$i\left(\frac{\partial\psi}{\partial t} + v_g\frac{\partial\psi}{\partial z}\right) + \frac{1}{2}D\frac{\partial^2\psi}{\partial z^2} - N|\psi|^2\psi = 0. \tag{9.62}$$

This is the *nonlinear Schrödinger (NLS) equation*. Now, consider a frame of reference moving at group velocity v_g, such that

$$z' = z - v_g t.$$

In this moving frame, the envelope is $u(z', t) = \psi(z, t)$, and (9.62) reduces to

$$\frac{1}{2}D\frac{\partial^2 u}{\partial z'^2} - N|u|^2 u = -i\frac{\partial u}{\partial t}. \tag{9.63}$$

Using $t = \frac{-tN}{2}$ and $z = z'\sqrt{\frac{-N}{D}}$, we reduce (9.63) to its canonical form

$$iu_t + u_{zz} + 2u^2 u^* = 0, \tag{9.64}$$

where the subscripts represent the partial derivatives.

When dispersion (D) and nonlinearity (N) compensate each other, we observe the formation of pulses known as solitons. To obtain the lowest order soliton, we substitute the shape preserving solution

$$u(z, t) = v(z) \exp[i\phi(z, t)] \tag{9.65}$$

Fig. 9.1 Shape of first and second order solitons. (Reproduced from [8] with permission.)

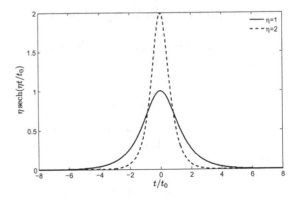

into (9.64), gather the real and imaginary parts and simplify them to get a second order differential equation

$$v_{zz} = v\eta^2 - 2v^3 \tag{9.66}$$

with a solution $v(z) = \eta \mathrm{sech}(\eta z)$ (Problem 9.8).

The general solution to (9.63) has an order parameter η, and is of the form

$$u(z, t) = \eta \, \mathrm{sech}(\eta z) \, \exp(i\eta^2 t). \tag{9.67}$$

Figure 9.1 is a plot of the first and second order solitons, where we observe that η determines both the soliton amplitude and its width.

Problem 9.6 Consider the dispersion relation

$$\left[k - k(\omega, |\psi|^2)\right]\psi = 0. \tag{9.68}$$

For a slowly varying amplitude ψ, a small deviation due to dispersion drives the wave number $k(\omega_0)$ to $k = k(\omega_0) + \delta k$.

(a) Using the Taylor expansion

$$\delta k = k_1 \delta\omega + \frac{k_2}{2}\delta\omega^2 + \frac{\partial k}{\partial|\psi|^2}|\psi|^2, \tag{9.69}$$

and identifying $t' = t - z/v_g$, derive the canonical NLS equation for a wave of the form $u(z, t)$ as

$$i\frac{\partial u}{\partial z} = \frac{k_2}{2}\frac{\partial^2 u}{\partial t^2} - \gamma|u|^2 u, \tag{9.70}$$

where k_2 and γ are the dispersion and nonlinearity coefficients, respectively.

(b) Verify, by substitution, that

$$u(z, t) = u_0 \text{sech}(t/t_0)e^{-i\kappa z} \tag{9.71}$$

is a solution to the NLS equation where the pulse amplitude u_0 and the width t_0 are related according to

$$|u_0|^2 = -\frac{k_2}{\gamma t_0^2} \quad \text{and} \quad \kappa = \frac{k_2}{2t_0^2}. \tag{9.72}$$

(c) A physical pulse must have $|u_0|^2 > 0$, i.e., $k_2\gamma < 0$. For forward volume spin waves, expand the dispersion relation of the lowest order mode for small kd and verify that this criterion is satisfied.

(d) Expand the dispersion relation of the lowest order forward volume wave mode for small kd and verify that $ND < 0$. Modes with $ND < 0$ are said to have *anomalous dispersion*.

Solution 9.6

(a) We begin with the dispersion relation

$$\left[k - k(\omega, |\psi|^2)\right]\psi = 0.$$

For small changes in wavenumber, we can expand k about ω_0 in a Taylor series, and write the dispersion relation as

$$\left[(k - k(\omega_0)) - \frac{\partial k}{\partial \omega}(\omega - \omega_0) - \frac{1}{2}\frac{\partial^2 k}{\partial \omega^2}(\omega - \omega_0)^2 - \frac{\partial k}{\partial |\psi|^2}\left(|\psi|^2 - 0\right)\right]\psi = 0 \tag{9.73}$$

Let $k(\omega_0) = k_0$, and affect the transformations

$$\omega - \omega_0 \to i\frac{\partial}{\partial t}, \quad (\omega - \omega_0)^2 \to -\frac{\partial^2}{\partial t^2} \quad \text{and} \quad (k - k_0) \to -i\frac{\partial}{\partial z}.$$

Observing that $\dfrac{\partial k}{\partial \omega} = \dfrac{1}{v_g}$, (9.73) becomes

$$-i\frac{\partial}{\partial z}\psi - \frac{i}{v_g}\frac{\partial}{\partial t}\psi + \frac{1}{2}\frac{\partial^2 k}{\partial \omega^2}\frac{\partial^2 \psi}{\partial t^2} - \frac{\partial k}{\partial |\psi|^2}|\psi|^2\psi = 0. \tag{9.74}$$

Now, let $t' = t - \dfrac{z}{v_g}$ and define $u(z, t') = \psi(z, t)$ such that

$$\frac{\partial \psi}{\partial z} = \frac{\partial u}{\partial z} + \frac{\partial u}{\partial t'}\frac{\partial t'}{\partial z} = \frac{\partial u}{\partial z} - \frac{1}{v_g}\frac{\partial u}{\partial t'}$$

and

$$\frac{\partial \psi}{\partial t} = \frac{\partial u}{\partial t'} \frac{\partial t'}{\partial t} = \frac{\partial u}{\partial t'}.$$

In writing the last equation, we observed that z and t are independent variables. Substituting these in (9.74), and dropping the primes with an overall negative sign,

$$i\frac{\partial u}{\partial z} - \frac{i}{v_g}\frac{\partial u}{\partial t} + \frac{i}{v_g}\frac{\partial u}{\partial t} - \frac{1}{2}\frac{\partial^2 k}{\partial \omega^2}\frac{\partial^2 u}{\partial t^2} + \frac{\partial k}{\partial |u|^2}|u|^2 u = 0$$

which becomes

$$\boxed{i\frac{\partial u}{\partial z} = \frac{k_2}{2}\frac{\partial^2 u}{\partial t^2} - \gamma |u|^2 u}$$

where

$$k_2 = \frac{\partial^2 k}{\partial \omega^2}, \quad \gamma = \frac{\partial k}{\partial |u|^2}.$$

(b) Given that

$$u = u_0 \mathrm{sech}\left(\frac{t}{t_0}\right) e^{-i\kappa z},$$

direct substitution yields

$$\frac{\partial u}{\partial z} = -i\kappa u \quad \text{or} \quad i\frac{\partial u}{\partial z} = \kappa u$$

and

$$\frac{\partial u}{\partial t} = -\frac{u_0}{t_0}\tanh\left(\frac{t}{t_0}\right)\mathrm{sech}\left(\frac{t}{t_0}\right)e^{-i\kappa z} = -\frac{u}{t_0}\tanh\left(\frac{t}{t_0}\right)$$

$$\frac{\partial^2 u}{\partial t^2} = -\frac{1}{t_0}\left[\tanh\left(\frac{t}{t_0}\right)\frac{\partial u}{\partial t} + \frac{u}{t_0}\mathrm{sech}^2\left(\frac{t}{t_0}\right)\right]$$

$$= \frac{u}{t_0^2}\left[\tanh^2\left(\frac{t}{t_0}\right) - \mathrm{sech}^2\left(\frac{t}{t_0}\right)\right].$$

Using the identity $1 - \tanh^2 \theta = \mathrm{sech}^2\theta$,

$$\frac{\partial^2 u}{\partial t^2} = \frac{u}{t_0^2}\left[1 - 2\mathrm{sech}^2\left(\frac{t}{t_0}\right)\right].$$

Hence

$$i\frac{\partial u}{\partial z} = \frac{k_2}{2}\frac{\partial^2 u}{\partial t^2} - \gamma |u|^2 u$$

$$\kappa u = \frac{k_2}{2}\frac{u}{t_0^2}\left[1 - 2\mathrm{sech}^2\left(\frac{t}{t_0}\right)\right] - \gamma |u_0|^2 \mathrm{sech}^2\left(\frac{t}{t_0}\right) u,$$

or

$$\frac{k_2}{2t_0^2} - \kappa - \left[\gamma|u_0|^2 + \frac{k_2}{t_0^2}\right] \text{sech}\left(\frac{t}{t_0}\right) = 0 \tag{9.75}$$

For (9.75) to be satisfied, we require that

$$\boxed{\kappa = \frac{k_2}{2t_0^2} \quad \text{and} \quad |u_0|^2 = -\frac{k_2}{\gamma t_0^2}.}$$

(c) First consider the nonlinear coefficient

$$\gamma = \frac{\partial k}{\partial|\psi|^2} = \frac{\partial k}{\partial|u|^2}.$$

(Note that the partial derivatives with respect to the magnitudes squared of ψ and u are equal, since these functions only differ by transformation of the coordinate system.) We can calculate γ with the aid of the chain rule

$$\frac{\partial k}{\partial|u|^2} = \frac{\partial k}{\partial\omega_0}\frac{\partial\omega_0}{\partial M_z}\frac{\partial M_z}{\partial|u|^2}. \tag{9.76}$$

To find $\partial k/\partial\omega_0$, let us use the approximation (5.45) for the lowest order forward volume wave mode

$$\omega^2 = \omega_0\left[\omega_0 + \omega_M\left(1 - \frac{1 - e^{-kd}}{kd}\right)\right]. \tag{9.77}$$

The behavior for small kd can be obtained using the Taylor/McLauren series expansion for the exponential:

$$e^{-kd} \approx 1 - kd + \frac{1}{2}(kd)^2 - \frac{1}{3!}(kd)^3.$$

Substituting this into (9.77) and simplifying gives

$$\omega^2 \approx \omega_0^2\left[1 + \frac{\omega_M}{\omega_0}\frac{kd}{2}\left(1 - \frac{kd}{3}\right)\right]. \tag{9.78}$$

Using the expansion $\sqrt{1 + x} \approx 1 + x/2$ for $|x| \ll 1$, the dispersion relation can be finally approximated as

$$\omega \approx \omega_0\left[1 + \frac{\omega_M}{\omega_0}\frac{kd}{4}\left(1 - \frac{kd}{3}\right)\right] = \omega_0 + \omega_M\frac{kd}{4} - \omega_M\frac{(kd)^2}{12}. \tag{9.79}$$

Keeping only terms linear in kd allows us to write

$$\omega \approx \omega_0 + \frac{\omega_M}{4}kd,$$

or

$$k \approx \frac{4}{\omega_M d}(\omega - \omega_0),$$

from which we readily obtain

$$\frac{\partial k}{\partial \omega_0} = -\frac{4}{\omega_M d}. \tag{9.80}$$

The next term in (9.76), $\partial \omega_0/\partial M_z$, is readily obtained from (see Problem 5.2, and Fig. 5.1)

$$\omega_0 = |\gamma_S|\mu_0(H_{DC} - M_z). \tag{9.81}$$

The result is

$$\frac{\partial \omega_0}{\partial M_z} = -|\gamma_S|\mu_0. \tag{9.82}$$

The final term in (9.76), $\partial M_z/\partial|u|^2$, can be obtained as follows. The z component of the magnetization is given by

$$M_z = M_S \cos\theta \approx M_S(1 - \theta^2/2). \tag{9.83}$$

Note that in writing (9.83) we are approximating the motion of the magnetization as circular and assuming the cone angle of the precession is small. Under these assumptions the small-signal magnetization can be expressed

$$m = M_S \sin\theta \approx M_S\theta. \tag{9.84}$$

The z component of the magnetization (9.83) can then be written

$$M_z \approx M_S \left[1 - \frac{1}{2}\left(\frac{m}{M_S}\right)^2\right] = M_S\left[1 - |\psi|^2\right] = M_S\left[1 - |u|^2\right], \tag{9.85}$$

where we have used the expression

$$|\psi|^2 = \frac{1}{2}\left(\frac{m}{M_S}\right)^2.$$

We can now readily obtain

$$\frac{\partial M_z}{\partial|u|^2} = -M_S. \tag{9.86}$$

We now have all the pieces to calculate γ by combining (9.80), (9.82), and (9.86)

$$\frac{\partial k}{\partial |u|^2} = \left(-\frac{4}{\omega_M d} \right) (-|\gamma_S| \mu_0) (-M_S)$$

$$= -\frac{4\omega_M}{\omega_M d}$$

$$= -\frac{4}{d}. \tag{9.87}$$

We conclude that

$$\boxed{\gamma = \frac{\partial k}{\partial |u|^2} = -\frac{4}{d} < 0.} \tag{9.88}$$

Next, we consider the sign of $k_2 = \partial^2 k / \partial \omega^2$. Taking the derivative with respect to k of (9.79) gives the group velocity:

$$\frac{\partial \omega}{\partial k} \approx \frac{\omega_M d}{4} - \frac{\omega_M k d^2}{6} = \frac{\omega_M d}{4} \left(1 - \frac{2kd}{3} \right), \tag{9.89}$$

or

$$\frac{\partial k}{\partial \omega} \approx \frac{4}{\omega_M d} \left(1 - \frac{2kd}{3} \right)^{-1}.$$

Taking the derivative with respect to ω gives

$$k_2 = \frac{\partial^2 k}{\partial \omega^2} \approx \frac{8}{3\omega_M} \left(1 - \frac{2kd}{3} \right)^{-2} \frac{\partial k}{\partial \omega} = \frac{32}{3\omega_M^2 d} \left(1 - \frac{2kd}{3} \right)^{-3}.$$

Thus, for $2kd/3 \ll 1$, it is clear that

$$\boxed{k_2 = \frac{\partial^2 k}{\partial \omega^2} \approx \frac{32}{3\omega_M^2 d} > 0.}$$

Combining this result with (9.88) gives the desired result

$$\boxed{k_2 \gamma < 0.}$$

(d) In (9.73), we used a Taylor series expansion of $k(\omega)$. Here, we pursue an equivalent approach with a series expansion of $\omega(k)$, with the dispersion coefficient $D = \partial^2 \omega / \partial k^2$ as defined in (9.61). Taking the derivative with respect to k of (9.89) gives

$$\boxed{D = \frac{\partial^2 \omega}{\partial k^2} \approx -\frac{\omega_M d^2}{6} < 0.} \tag{9.90}$$

To find the nonlinearity coefficient N let us return to the dispersion relation (9.79), take the limit as $kd \to 0$, and expand ω_0 using (9.81) and (9.85). The result is

$$\omega \approx \omega_{DC} + \omega_M \left(|\psi|^2 - 1\right).$$

The nonlinearity coefficient can now be written

$$N = \frac{\partial \omega}{\partial(\psi^2)} \approx \omega_M > 0. \tag{9.91}$$

Combining (9.90) and (9.91) gives

$$ND = -\frac{\omega_M^2 d^2}{6} < 0.$$

We conclude that the conditions of anomalous dispersion are satisfied.

Problem 9.7 (a) Using appropriate hyperbolic trigonometric identities, show by direct substitution that

$$u(z, t) = \eta \operatorname{sech}(\eta z) \exp(i\eta^2 t) \tag{9.92}$$

is a solution to the nonlinear Schrödinger equation (9.64)

$$i u_t + u_{zz} + 2u^2 u^* = 0 \tag{9.93}$$

(b) If $u(z, t)$ is a solution to the NLS equation, show that $\varepsilon u(\varepsilon z, \varepsilon^2 t)$ is also a solution. The arbitrary scaling factor ε accounts for the existence of higher order solitons, with the fundamental soliton obtained by choosing $u(0, 0) = 1$ and $\varepsilon = 1$.

Solution 9.7

(a) We begin with the trigonometric identities

$$\frac{d}{d\theta} \operatorname{sech}\theta = -\tanh\theta \operatorname{sech}\theta$$

$$\frac{d^2}{d\theta^2} \operatorname{sech}\theta = \frac{d}{d\theta}[-\tanh\theta \operatorname{sech}\theta]$$

$$= \tanh^2\theta \operatorname{sech}\theta - \operatorname{sech}\theta \frac{d}{d\theta}\tanh\theta$$

Now

$$\frac{d}{d\theta}\tanh\theta = \operatorname{sech}^2\theta.$$

Hence

$$\frac{d^2}{d\theta^2}\text{sech}\theta = \tanh^2\theta\,\text{sech}\theta - \text{sech}^3\theta = \text{sech}\theta\left(\tanh^2\theta - \text{sech}^2\theta\right).$$

Since $\tanh^2\theta = 1 - \text{sech}^2\theta$, we have

$$\frac{d^2}{d\theta^2}\text{sech}\theta = \text{sech}\theta\left(1 - 2\text{sech}^2\theta\right) = \text{sech}\theta - 2\text{sech}^3\theta.$$

Now let $u(z,t) = \eta\,\text{sech}(\eta z)e^{i\eta^2 t}$. By substitution, we observe that

$$
\begin{aligned}
iu_t + u_{zz} + 2|u|^2 u &= -\eta^3\text{sech}(\eta z)e^{i\eta^2 t} + \eta^3\left(\text{sech}(\eta z) - 2\text{sech}^3(\eta z)\right)e^{i\eta^2 t} \\
&\quad + 2\eta^3\text{sech}^3(\eta t)e^{i\eta^2 t} \\
&= 0.
\end{aligned}
$$

Thus

$$\boxed{u(z,t) = \eta\,\text{sech}(\eta z)e^{i\eta^2 t} \quad \text{is a solution to} \quad iu_t + u_{zz} + 2|u|^2 u.}$$

(b) We begin with the assumption that $u(z,t)$ is a solution to the NLS equation. Differentiating $u(\varepsilon z, \varepsilon^2 t)$ with respect to t and z yields,

$$u_t(\varepsilon z, \varepsilon^2 t) = \varepsilon^2 u_\tau(\lambda, \tau),$$
$$u_{zz}(\varepsilon z, \varepsilon^2 t) = \varepsilon^2 u_{\lambda\lambda}(\lambda, \tau).$$

Here $\lambda = \varepsilon z$, and $\tau = \varepsilon^2 t$. Substituting $v(z,t) = \varepsilon u(\varepsilon z, \varepsilon^2 t)$ into the NLS equation, we get

$$
\begin{aligned}
iv_t + v_{zz} + 2|v|^2 v &= i\varepsilon u_t(\varepsilon z, \varepsilon^2 t) + \varepsilon u_{zz}(\varepsilon z, \varepsilon^2 t) + 2|\varepsilon u(\varepsilon z, \varepsilon^2 t)|^2 \varepsilon u(\varepsilon z, \varepsilon^2 t) \\
&= \varepsilon^3\left[iu_\tau(\lambda, \tau) + u_{\lambda\lambda}(\lambda, \tau) + 2|u(\lambda, \tau)|^2 u(\lambda, \tau)\right] \\
&= 0
\end{aligned}
$$

thus proving that $v(z,t) = \varepsilon u(\varepsilon z, \varepsilon^2 t)$ is also a solution.

Problem 9.8 Multiply (9.66) by $2(dv/dz)$ and integrate both sides to obtain

$$v_z = \left[v^2\eta^2 - v^4 + C\right]^{1/2}, \tag{9.94}$$

where C is a constant of integration. Requiring that both v and v_z vanish as $z \to \infty$ yields $C = 0$. Using $v = 1/s$ and $s = \eta^{-1}\cosh\theta$, integrate (9.94) to obtain $\theta = -\eta z$ and thus, $v(z) = \eta\,\text{sech}(\eta z)$.

Solution 9.8

From (9.66), we have

$$v_{zz} = v\eta^2 - 2v^3.$$

Multiplying both sides by $2dv/dz = 2v_z$ gives

$$2v_{zz}v_z = 2\left(v\eta^2 - 2v^3\right)v_z. \tag{9.95}$$

Note that

$$\frac{d}{dz}(v_z)^2 = 2v_z\frac{dv_z}{dz} = 2v_zv_{zz}.$$

Consequently, (9.95) can be written

$$\frac{d}{dz}(v_z)^2 = 2\left(v\eta^2 - 2v^3\right)\frac{dv}{dz}.$$

Integrating both sides gives

$$v_z^2 = 2\left(\frac{v^2}{2}\eta^2 - \frac{2v^4}{4}\right) + C,$$

$$v_z = \sqrt{v^2\eta^2 - v^4 + C}.$$

Since $v, v_z \to 0$ as $z \to \infty$, we conclude $C = 0$. Let $v = 1/s$

$$\frac{d}{dz}\left(\frac{1}{s}\right) = \left[\frac{\eta^2}{s^2} - \frac{1}{s^4}\right]^{1/2}$$

$$-\frac{1}{s^2}\frac{ds}{dz} = \left[\frac{\eta^2}{s^2} - \frac{1}{s^4}\right]^{1/2}$$

$$\therefore \quad -\frac{ds}{dz} = \left[s^2\eta^2 - 1\right]^{1/2}$$

Next, let $s = \frac{1}{\eta}\cosh\theta$

$$-\frac{1}{\eta}\sinh\theta\frac{d\theta}{dz} = \left[\frac{\eta^2\cosh^2\theta}{\eta^2} - 1\right]^{1/2}$$

$$= \left[\cosh^2\theta - 1\right]^{1/2}$$

$$= \sinh\theta$$

which reduces to

$$\frac{d\theta}{dz} = -\eta,$$

or

$$\theta = -\eta z + C.$$

However, the integration constant simply corresponds to an offset in position of the pulse, and can be set to zero without loss of generality. We conclude that

$$s = \frac{1}{\eta} \cosh(-\eta z) = \frac{1}{\eta} \cosh(\eta z)$$

$$\therefore \quad \boxed{v = \frac{1}{s} = \eta \operatorname{sech}(\eta z)}$$

References

1. H. Suhl, The theory of ferromagnetic resonance at high signal powers. J. Phys. Chem. Solids **1**, 209 (1957)
2. E. Schlömann, Ferromagnetic resonance at high power levels. Technical Report (Raytheon Corporation, Waltham, 1959)
3. P. Krivosik, C.E. Patton, Hamiltonian formulation of nonlinear spin-wave dynamics: theory and applications. Phys. Rev. B **82**, 184428 (2010)
4. V.E. Zakharov, V.S. L'vov, S.S. Starobinets, Instability of monochromatic spin waves. Sov. Phys. Solid State **11**, 2368 (1970)
5. M.J. Ablowitz, P.A. Clarkson, *Solitons, Nonlinear Evolution Equations and Inverse Scattering* (Cambridge University Press, New York, 1991)
6. P. Wigen (ed.), *Nonlinear Phenomena and Chaos in Magnetic Materials* (World Scientific, Singapore, 1994)
7. P. Krivosik, N. Mo, S. Kalarickal, C.E. Patton, Hamiltonian formalism for two magnon scattering microwave relaxation: theory and applications. J. App. Phys. **101**, 083901 (2007)
8. D.D. Stancil, A. Prabhakar, *Spin Waves: Theory and Applications* (Springer, New York, 2009)

Novel Applications

10

The introduction of a magnesium oxide insulating layer, within a magnetic tunnel junction, brought about a significant increase in the sensitivity of tunnel magnetoresistive devices. Alongside this development, we also began observing a spin transfer torque (STT). Today, STT based devices form the basis of a new generation of random access memory devices, namely the STT-RAMs. Spin current injection via nano-contacts can also be used to excite spin waves [1], but it requires that we efficiently mode-match with the guided spin wave modes. In this chapter, we explore nano-contact spin wave generation structures using current-driven spin transfer torque, and the "left-handed" behavior of backward volume spin waves leading to the observation of the inverse Doppler effect.

Problems 10.1–10.3 are concerned with spin rotation and spin current calculations. Problem 10.1 establishes the effects of Pauli spin matrices operating on spinors (vectors representing spin states), and echoes some of the ideas first introduced in Chaps. 1 and 2. Problems 10.2 and 10.3 then build on the relations established in Problem 10.1 to fill in some of the derivation steps for the spin torque expressions. Problem 10.4 shifts the topic to left-handed spin waves, and shows that backward volume spin waves have a net left-handed behavior, although the contributions from the fields inside and outside of the magnetic layer are in opposition.

10.1 Spin Transfer Torque

The excitation of spin waves by a DC current was independently predicted by Slonczewski [2] and Berger [3], and has received significant attention as a novel way to generate microwave signals. Here we give a brief simple model for the effect based on the analysis of Stiles and Zangwill [4].

Consider the multilayer geometry shown in Fig. 10.1. In a ferromagnetic metal, electrons with spin-up with respect to the effective molecular field will have their energy shifted compared to those electrons with spin-down. As a result, the spin-up and spin-down energy bands will be shifted so that when the bands are filled up to the Fermi level, there will be more electrons in the spin-down band. (Recall that since the charge is negative, the magnetic moment is opposite to the spin. Consequently, the lowest value of Zeeman energy $-\boldsymbol{\mu} \cdot \mathbf{H}_m$ occurs when $\boldsymbol{\mu}$ is parallel to $\mathbf{H}_m$, or when $\mathbf{S}$ is antiparallel to $\mathbf{H}_m$.)

As a result, the current injected into layer F$_2$ from F$_1$ is *spin polarized*, i.e., there are more electrons with spin parallel to $\mathbf{S}_1$ than antiparallel to $\mathbf{S}_1$. Referring to the dashed box in Fig. 10.1, if the angular momentum associated with $\mathbf{J}_{F_2}$ is different from the angular momentum associated with $\mathbf{J}_N$, then there must be a torque on the spins in the dashed box to conserve angular momentum, provided there are no other torques acting on the system and there is no spin accumulation in the region Δx. Consequently, we can gain some understanding of the origin of the spin transfer torque on $\mathbf{S}_2$ by just considering the single interface between the normal metal N and the ferromagnetic region F$_2$. For simplicity, we will model the electron wave functions as uniform plane waves.

The transmission and reflection of plane waves at a planar interface is a standard problem in quantum mechanics (e.g., see [5] or [6]). The transmission and reflection coefficients for electrons with spin σ are given by

$$r_\sigma = \frac{k_N - k_\sigma}{k_N + k_\sigma}, \tag{10.1}$$

$$t_\sigma = \frac{2\sqrt{k_N k_\sigma}}{k_N + k_\sigma}, \tag{10.2}$$

and

$$r_\sigma{}^2 + t_\sigma{}^2 = 1. \tag{10.3}$$

Here k_N is the wave number for electrons in the normal metal layer, and k_σ is the wave number for electrons in layer F$_2$ with spin σ.

In terms of these transmission and reflection coefficients, plane wave expressions for the incident, reflected, and transmitted wave functions for an electron with spin oriented at the angle θ with respect to the z axis can be written

$$\psi_{\text{in}} = \frac{e^{ikx}}{\sqrt{k}} \left(\cos(\theta/2)\, \chi_\uparrow + \sin(\theta/2)\, \chi_\downarrow \right), \tag{10.4a}$$

$$\psi_{\text{ref}} = \frac{e^{-ikx}}{\sqrt{k}} \left(r_\uparrow \cos(\theta/2)\, \chi_\uparrow + r_\downarrow \sin(\theta/2)\, \chi_\downarrow \right), \tag{10.4b}$$

$$\psi_{\text{tr}} = t_\uparrow \cos(\theta/2)\, \frac{e^{ik_\uparrow x}}{\sqrt{k_\uparrow}} \chi_\uparrow + t_\downarrow \sin(\theta/2)\, \frac{e^{ik_\downarrow x}}{\sqrt{k_\downarrow}} \chi_\downarrow. \tag{10.4c}$$

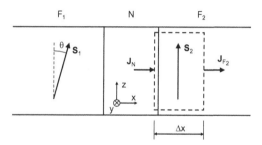

Fig. 10.1 Multilayer structure consists of two ferromagnetic metal layers separated by a thin normal metal layer. A current is injected into the structure from the left, resulting in a spin polarized current injected in the layer F_2 (From [6], reproduced with permission)

To proceed with the analysis, we also define the *spin density* **S** and *spin current density* **J** as

$$
\mathbf{S} = \frac{\hbar}{2}\psi^\dagger \boldsymbol{\sigma} \psi, \tag{10.5}
$$

$$
\mathbf{J} = \frac{\hbar^2}{2m}\,\mathrm{Im}(\psi^\dagger \boldsymbol{\sigma} \otimes \nabla\psi), \tag{10.6}
$$

where $\boldsymbol{\sigma}$ is the vector of Pauli matrices (see Problems 1.5 and 10.1). Here $\otimes$ represents the tensor product. (The wave functions are normalized so that the particle flux $\psi^\dagger \nabla\psi = 1$ for our 1-dimensional problem.) Note that **J** is a tensor whose left index represents the spin component, and whose right index represents the direction of particle travel.

With current along the x direction as in Fig. 10.1, time rate of change of the spin density $\mathbf{S}_2$ is proportional to the net spin torque density per unit area **N**:

$$
\frac{\partial \mathbf{S}_2}{\partial t}\Delta x = \left[\mathbf{J}^{\mathrm{in}} + \mathbf{J}^{\mathrm{ref}} - \mathbf{J}^{\mathrm{tr}}\right]\cdot\hat{\mathbf{x}} \equiv \mathbf{N}. \tag{10.7}
$$

Using (10.4) for the wave functions and (10.6) for the spin current density, the net torque per unit area can be written as (see Problem 10.1 for the details)

$$
\mathbf{N} = \frac{\hbar^2 \sin\theta}{2m}\left\{\hat{\mathbf{x}}\left[1 - \mathrm{Re}\left(r_\uparrow^* r_\downarrow + \frac{(k_\uparrow + k_\downarrow)}{2\sqrt{k_\uparrow k_\downarrow}}t_\uparrow^* t_\downarrow e^{i(k_\downarrow - k_\uparrow)x}\right)\right]\right.
$$
$$
\left. - \hat{\mathbf{y}}\,\mathrm{Im}\left(r_\uparrow^* r_\downarrow + \frac{(k_\uparrow + k_\downarrow)}{2\sqrt{k_\uparrow k_\downarrow}}t_\uparrow^* t_\downarrow e^{i(k_\downarrow - k_\uparrow)x}\right)\right\}. \tag{10.8}
$$

We have written this expression for particular values of $k_\uparrow$ and $k_\downarrow$, but in reality electrons at the Fermi surface may have a range of values depending on the transverse component of the electron wave vector q. Specifically, we have

$$
k_{\mathrm{N}} = \sqrt{k_F^2 - q^2}, \quad k_\uparrow = \sqrt{k_{F\uparrow}^2 - q^2}, \quad \text{and} \quad k_\downarrow = \sqrt{k_{F\downarrow}^2 - q^2}. \tag{10.9}
$$

Note that although the wave numbers at the Fermi surface k_F, $k_{F\uparrow}$, and $k_{F\downarrow}$ are in general different, the boundary conditions require the component of the wave vector parallel to the boundary surface to be continuous. Hence, q is the same for each, and takes on the range of values appropriate for the normal metal since the incident current is in this region, i.e., $0 \leq q \leq k_F$.

Following [4], we note that

$$r_{\uparrow}^* r_{\downarrow} = |r_{\uparrow}^* r_{\downarrow}| e^{i \Delta \phi(q)}, \tag{10.10}$$

where $\Delta \phi(q)$ is a phase angle depending on the angle of incidence on the interface. If the variation of $\Delta \phi$ over the Fermi surface is sufficiently large, this term will tend to average to zero so that

$$\langle r_{\uparrow}^* r_{\downarrow} \rangle \approx 0. \tag{10.11}$$

Following a similar argument

$$\left\langle \frac{(k_{\uparrow} + k_{\downarrow})}{2\sqrt{k_{\uparrow} k_{\downarrow}}} t_{\uparrow}^* t_{\downarrow} e^{i(k_{\downarrow} - k_{\uparrow})x} \right\rangle \approx 0 \tag{10.12}$$

owing to the oscillations in $t_{\uparrow}^* t_{\downarrow} \exp(i(k_{\downarrow} - k_{\uparrow})x)$ when averaging over the Fermi surface.

Consequently, the net torque per unit area transferred into the region Δx is approximated by

$$\mathbf{N} = \hat{\mathbf{x}} \frac{\hbar^2}{2m} \sin \theta. \tag{10.13}$$

Noting that $\hat{\mathbf{S}}_2 \times (\hat{\mathbf{S}}_1 \times \hat{\mathbf{S}}_2) = \hat{\mathbf{x}} \sin \theta$, the spin transfer torque can be written as

$$\frac{\partial \mathbf{S}_2}{\partial t} \Delta x = \frac{\hbar^2}{2m} \hat{\mathbf{S}}_2 \times (\hat{\mathbf{S}}_1 \times \hat{\mathbf{S}}_2). \tag{10.14}$$

If layer 2 is a thin ferromagnetic layer of thickness d, then the spin transfer torque can be finally written

$$\frac{\partial \mathbf{S}_2}{\partial t} = \frac{\hbar^2}{2md} \hat{\mathbf{S}}_2 \times (\hat{\mathbf{S}}_1 \times \hat{\mathbf{S}}_2), \tag{10.15}$$

where we have set $\Delta x = d$. While damping results from a torque driving a precessing spin toward its equilibrium ground state, the spin transfer torque is in the direction to increase the precession angle and can be used to overcome loss resulting in oscillation.

Problem 10.1 The Pauli matrices $\boldsymbol{\sigma}$, introduced in Problem 1.5, act as rotation operators on the *spinor* state. A rotation by an angle θ about the axis $\hat{\mathbf{n}}$ is written as [5]

$$U_R = \mathbf{I} \cos \frac{\theta}{2} - i\hat{\mathbf{n}} \cdot \boldsymbol{\sigma} \sin \frac{\theta}{2}, \tag{10.16}$$

where $\mathbf{I}$ is the identity matrix and

$$\sigma_1 = \begin{bmatrix} 0 & 1 \\ 1 & 0 \end{bmatrix}, \quad \sigma_2 = \begin{bmatrix} 0 & -i \\ i & 0 \end{bmatrix}, \quad \sigma_3 = \begin{bmatrix} 1 & 0 \\ 0 & -1 \end{bmatrix}. \tag{10.17}$$

Using the notation

$$\chi_\uparrow = \begin{bmatrix} 1 \\ 0 \end{bmatrix}; \quad \chi_\downarrow = \begin{bmatrix} 0 \\ 1 \end{bmatrix}, \tag{10.18}$$

show that

$$\sigma_1 \chi_\uparrow = \chi_\downarrow, \tag{10.19}$$

$$i\sigma_2 \chi_\uparrow = -\chi_\downarrow, \tag{10.20}$$

$$\sigma_3 \chi_\uparrow = \chi_\uparrow, \tag{10.21}$$

and derive the equivalent operations on $\chi_\downarrow$.

Solution 10.1

Using the Pauli spin matrices and spinors given in (10.17) and (10.18), we can calculate the effects of the Pauli matrix operators as follows:

$$\sigma_1 \chi_\uparrow = \begin{bmatrix} 0 & 1 \\ 1 & 0 \end{bmatrix} \begin{bmatrix} 1 \\ 0 \end{bmatrix} = \begin{bmatrix} 0 \\ 1 \end{bmatrix} = \chi_\downarrow$$

$$i\sigma_2 \chi_\uparrow = i \begin{bmatrix} 0 & -i \\ i & 0 \end{bmatrix} \begin{bmatrix} 1 \\ 0 \end{bmatrix} = \begin{bmatrix} 0 \\ -1 \end{bmatrix} = -\chi_\downarrow$$

$$\sigma_3 \chi_\uparrow = \begin{bmatrix} 1 & 0 \\ 0 & -1 \end{bmatrix} \begin{bmatrix} 1 \\ 0 \end{bmatrix} = \begin{bmatrix} 1 \\ 0 \end{bmatrix} = \chi_\uparrow$$

Similarly

$$\sigma_1 \chi_\downarrow = \begin{bmatrix} 0 & 1 \\ 1 & 0 \end{bmatrix} \begin{bmatrix} 0 \\ 1 \end{bmatrix} = \begin{bmatrix} 1 \\ 0 \end{bmatrix} = \chi_\uparrow$$

$$i\sigma_2 \chi_\downarrow = i \begin{bmatrix} 0 & -i \\ i & 0 \end{bmatrix} \begin{bmatrix} 0 \\ 1 \end{bmatrix} = \begin{bmatrix} 1 \\ 0 \end{bmatrix} = \chi_\uparrow$$

$$\sigma_3 \chi_\downarrow = \begin{bmatrix} 1 & 0 \\ 0 & -1 \end{bmatrix} \begin{bmatrix} 0 \\ 1 \end{bmatrix} = \begin{bmatrix} 0 \\ -1 \end{bmatrix} = -\chi_\downarrow.$$

To summarize

$\sigma_1 \chi_\uparrow = \chi_\downarrow,$	$\sigma_1 \chi_\downarrow = \chi_\uparrow$
$i\sigma_2 \chi_\uparrow = -\chi_\downarrow,$	$i\sigma_2 \chi_\downarrow = \chi_\uparrow$
$\sigma_3 \chi_\uparrow = \chi_\uparrow,$	$\sigma_3 \chi_\downarrow = -\chi_\downarrow$

Problem 10.2 Show that substituting incident, reflected, and transmitted wave functions given by (10.4) into

$$\mathbf{J} = \frac{\hbar^2}{2m} \text{Im}(\psi^\dagger \boldsymbol{\sigma} \otimes \nabla \psi), \tag{10.22}$$

yields

$$\mathbf{J}^{\text{in}} \cdot \hat{\mathbf{x}} = \frac{\hbar^2}{2m} \left[\hat{\mathbf{x}} \sin\theta + \hat{\mathbf{z}} \cos\theta \right], \tag{10.23a}$$

$$\mathbf{J}^{\text{ref}} \cdot \hat{\mathbf{x}} = \frac{\hbar^2}{2m} \left[-\hat{\mathbf{x}} \text{Re}(r_\uparrow^* r_\downarrow) \sin\theta - \hat{\mathbf{y}} \text{Im}(r_\uparrow^* r_\downarrow) \sin\theta \right.$$
$$\left. - \hat{\mathbf{z}} \left(|r_\uparrow|^2 \cos^2(\theta/2) - |r_\downarrow|^2 \sin^2(\theta/2) \right) \right], \tag{10.23b}$$

$$\mathbf{J}^{\text{tr}} \cdot \hat{\mathbf{x}} = \frac{\hbar^2}{2m} \left[\hat{\mathbf{x}} \left(\frac{k_\downarrow + k_\uparrow}{2\sqrt{k_\downarrow k_\uparrow}} \right) \text{Re}(t_\uparrow^* t_\downarrow e^{i(k_\downarrow - k_\uparrow)x}) \sin\theta \right.$$
$$+ \hat{\mathbf{y}} \left(\frac{k_\downarrow + k_\uparrow}{2\sqrt{k_\downarrow k_\uparrow}} \right) \text{Im}(t_\uparrow^* t_\downarrow e^{i(k_\downarrow - k_\uparrow)x}) \sin\theta$$
$$\left. + \hat{\mathbf{z}} \left(|t_\uparrow|^2 \cos^2(\theta/2) - |t_\downarrow|^2 \sin^2(\theta/2) \right) \right]. \tag{10.23c}$$

Solution 10.2

The general expression for spin current density is given by (10.6)

$$\overline{J} = \frac{\hbar^2}{2m} \text{Im}(\psi^\dagger \overline{\sigma} \otimes \nabla \psi).$$

The x component of current is

$$\overline{J} \cdot \hat{x} = \frac{\hbar^2}{2m} \text{Im}(\psi^\dagger \overline{\sigma} \otimes \frac{\partial \psi}{\partial x})$$
$$= \hat{x} J_{xx} + \hat{y} J_{yx} + \hat{z} J_{zx}.$$

To facilitate the computation, let us introduce the complex spin current $\tilde{J}_{ij}$ defined such that

$$J_{ij} = \frac{\hbar^2}{2m} \text{Im}(\tilde{J}_{ij}). \tag{10.24}$$

Let us first consider $\tilde{J}^{in}$, using ψ_{in} from (10.4a):

$$\tilde{J}^{in}_{xx} = \psi^\dagger_{in}\sigma_1 \frac{\partial \psi_{in}}{\partial x}$$

$$= \frac{e^{-ikx}}{\sqrt{k}} \left(\cos(\theta/2)\,\chi^\dagger_\uparrow + \sin(\theta/2)\,\chi^\dagger_\downarrow\right)\sigma_1 ik\frac{e^{ikx}}{\sqrt{k}}\left(\cos(\theta/2)\,\chi_\uparrow + \sin(\theta/2)\,\chi_\downarrow\right)$$

$$= i\left[\cos^2(\theta/2)\,\chi^\dagger_\uparrow\sigma_1\chi_\uparrow + \cos(\theta/2)\sin(\theta/2)\left(\chi^\dagger_\uparrow\sigma_1\chi_\downarrow + \chi^\dagger_\downarrow\sigma_1\chi_\uparrow\right)\right.$$

$$\left. + \sin^2(\theta/2)\,\chi^\dagger_\downarrow\sigma_1\chi_\downarrow\right].$$

$$(10.25)$$

Clearly we are going to need to calculate quantities of the form $\chi^\dagger\sigma_i\chi$. Using the results of Problem 10.1 along with the orthogonality property $\chi^\dagger_\uparrow\chi_\downarrow = \chi^\dagger_\downarrow\chi_\uparrow = 0$ leads to:

$$\begin{array}{ll} \chi^\dagger_\uparrow\sigma_1\chi_\uparrow = 0 & \chi^\dagger_\downarrow\sigma_1\chi_\uparrow = 1 \\ \chi^\dagger_\uparrow\sigma_2\chi_\uparrow = 0 & \chi^\dagger_\downarrow\sigma_2\chi_\uparrow = i \\ \chi^\dagger_\uparrow\sigma_3\chi_\uparrow = 1 & \chi^\dagger_\downarrow\sigma_3\chi_\uparrow = 0 \end{array}$$

$$\begin{array}{ll} \chi^\dagger_\uparrow\sigma_1\chi_\downarrow = 1 & \chi^\dagger_\downarrow\sigma_1\chi_\downarrow = 0 \\ \chi^\dagger_\uparrow\sigma_2\chi_\downarrow = -i & \chi^\dagger_\downarrow\sigma_2\chi_\downarrow = 0 \\ \chi^\dagger_\uparrow\sigma_3\chi_\downarrow = 0 & \chi^\dagger_\downarrow\sigma_3\chi_\downarrow = -1 \end{array}$$

Using these results, we have

$$\tilde{J}^{in}_{xx} = 2i\cos(\theta/2)\sin(\theta/2) = i\sin\theta.$$

In the last step we have used the identity $\sin(2\theta) = 2\cos\theta\sin\theta$. The xx component of the spin current tensor is, therefore

$$\boxed{J^{in}_{xx} = \frac{\hbar^2}{2m}\sin\theta.}$$

The components J^{in}_{yx}, J^{in}_{zx} can be obtained from (10.25) by replacing σ_1 by σ_2 and σ_3, respectively. Thus

$$\tilde{J}^{in}_{yx} = i\left[\cos^2(\theta/2)\,\chi^\dagger_\uparrow\sigma_2\chi_\uparrow + \cos(\theta/2)\sin(\theta/2)\left(\chi^\dagger_\uparrow\sigma_2\chi_\downarrow + \chi^\dagger_\downarrow\sigma_2\chi_\uparrow\right)\right.$$

$$\left. + \sin^2(\theta/2)\,\chi^\dagger_\downarrow\sigma_2\chi_\downarrow\right]$$

$$= i\cos(\theta/2)\sin(\theta/2)(-i+i)$$

$$= 0,$$

and

$$\boxed{J^{in}_{yx} = 0.}$$

Similarly

$$\tilde{J}_{zx}^{\text{in}} = i\left[\cos^2(\theta/2)\,\chi_\uparrow^\dagger\sigma_3\chi_\uparrow + \cos(\theta/2)\sin(\theta/2)\left(\chi_\uparrow^\dagger\sigma_3\chi_\downarrow + \chi_\downarrow^\dagger\sigma_3\chi_\uparrow\right)\right.$$
$$\left. + \sin^2(\theta/2)\,\chi_\downarrow^\dagger\sigma_3\chi_\downarrow\right]$$
$$= i\left[\cos^2(\theta/2) - \sin^2(\theta/2)\right]$$
$$= i\cos\theta,$$

where we have used the identity $\cos^2\theta - \sin^2\theta = \cos(2\theta)$. Thus

$$\boxed{J_{zx}^{\text{in}} = \frac{\hbar^2}{2m}\cos\theta,}$$

and

$$\boxed{\overline{J}^{\text{in}}\cdot\hat{x} = \frac{\hbar^2}{2m}\left(\hat{x}\sin\theta + \hat{z}\cos\theta\right).}$$

Next, we consider $\overline{J}^{\text{ref}}\cdot\hat{x}$, using ψ_{ref} from (10.4b)

$$\tilde{J}_{xx}^{\text{ref}} = \psi_{\text{ref}}^\dagger\sigma_1\frac{\partial\psi_{\text{ref}}}{\partial x}$$
$$= \frac{e^{ikx}}{\sqrt{k}}\left(r_\uparrow^*\cos(\theta/2)\,\chi_\uparrow^\dagger + r_\downarrow^*\sin(\theta/2)\,\chi_\downarrow^\dagger\right)\sigma_1(-ik)$$
$$\times\frac{e^{-ikx}}{\sqrt{k}}\left(r_\uparrow\cos(\theta/2)\,\chi_\uparrow + r_\downarrow\sin(\theta/2)\,\chi_\downarrow\right)$$
$$= -i\left[|r_\uparrow|^2\cos^2(\theta/2)\,\chi_\uparrow^\dagger\sigma_1\chi_\uparrow + \cos(\theta/2)\sin(\theta/2)\right.$$
$$\times\left(r_\uparrow^*r_\downarrow\chi_\uparrow^\dagger\sigma_1\chi_\downarrow + r_\downarrow^*r_\uparrow\chi_\downarrow^\dagger\sigma_1\chi_\uparrow\right) + |r_\downarrow|^2\sin^2(\theta/2)\,\chi_\downarrow^\dagger\sigma_1\chi_\downarrow\right]$$
$$= -i\cos(\theta/2)\sin(\theta/2)\left(r_\uparrow^*r_\downarrow + r_\downarrow^*r_\uparrow\right)$$
$$= -i2\cos(\theta/2)\sin(\theta/2)\,\text{Re}\left(r_\uparrow^*r_\downarrow\right)$$
$$= -i\sin\theta\,\text{Re}\left(r_\uparrow^*r_\downarrow\right),$$

and

$$\boxed{J_{xx}^{\text{ref}} = -\frac{\hbar^2}{2m}\sin\theta\,\text{Re}\left(r_\uparrow^*r_\downarrow\right).}$$

Similarly, from (10.4)

$$\tilde{J}^{\text{ref}}_{yx} = -i \left[|r_\uparrow|^2 \cos^2 (\theta/2) \, \chi^\dagger_\uparrow \sigma_2 \chi_\uparrow + \cos (\theta/2) \sin (\theta/2) \right.$$

$$\times \left(r^*_\uparrow r_\downarrow \chi^\dagger_\uparrow \sigma_2 \chi_\downarrow + r^*_\downarrow r_\uparrow \chi^\dagger_\downarrow \sigma_2 \chi_\uparrow \right) + |r_\downarrow|^2 \sin^2 (\theta/2) \, \chi^\dagger_\downarrow \sigma_2 \chi_\downarrow \right]$$

$$= -i \cos (\theta/2) \sin (\theta/2) \left(-i r^*_\uparrow r_\downarrow + i r^*_\downarrow r_\uparrow \right)$$

$$= - \cos (\theta/2) \sin (\theta/2) \left(r^*_\uparrow r_\downarrow - r^*_\downarrow r_\uparrow \right)$$

$$= -i 2 \cos (\theta/2) \sin (\theta/2) \, \text{Im} \left(r^*_\uparrow r_\downarrow \right)$$

$$= -i \sin \theta \, \text{Im} \left(r^*_\uparrow r_\downarrow \right),$$

and

$$\boxed{J^{\text{ref}}_{yx} = -\frac{\hbar^2}{2m} \sin \theta \, \text{Im} \left(r^*_\uparrow r_\downarrow \right).}$$

Calculating the last element of the reflected current tensor gives

$$\tilde{J}^{\text{ref}}_{zx} = -i \left[|r_\uparrow|^2 \cos^2 (\theta/2) - |r_\downarrow|^2 \sin^2 (\theta/2) \right],$$

and

$$\boxed{J^{\text{ref}}_{zx} = -\frac{\hbar^2}{2m} \left[|r_\uparrow|^2 \cos^2 \left(\frac{\theta}{2} \right) - |r_\downarrow|^2 \sin^2 \left(\frac{\theta}{2} \right) \right].}$$

Putting the pieces together gives the x component of the reflected spin current

$$\boxed{\begin{aligned} \overline{J}^{\text{ref}} \cdot \hat{x} = -\frac{\hbar^2}{2m} &\left[\hat{x} \sin \theta \, \text{Re} \left(r^*_\uparrow r_\downarrow \right) + \hat{y} \sin \theta \, \text{Im} \left(r^*_\uparrow r_\downarrow \right) \right. \\ &\left. + \hat{z} \left(|r_\uparrow|^2 \cos^2 \left(\frac{\theta}{2} \right) - |r_\downarrow|^2 \sin^2 \left(\frac{\theta}{2} \right) \right) \right]. \end{aligned}}$$

Finally, we consider $\overline{J}^{\text{tr}} \cdot \hat{x}$, using ψ_{tr} from (10.4c)

$$\tilde{J}^{\text{tr}}_{xx} = \psi^\dagger_{\text{tr}} \sigma_1 \frac{\partial \psi_{\text{tr}}}{\partial x}$$

$$= \left(t^*_\uparrow \cos\left(\frac{\theta}{2}\right) \frac{e^{-ik_\uparrow x}}{\sqrt{k_\uparrow}} \chi^\dagger_\uparrow + t^*_\downarrow \sin\left(\frac{\theta}{2}\right) \frac{e^{-ik_\downarrow x}}{\sqrt{k_\downarrow}} \chi^\dagger_\downarrow \right) \sigma_1 \left(ik_\uparrow t_\uparrow \cos\left(\frac{\theta}{2}\right) \frac{e^{ik_\uparrow x}}{\sqrt{k_\uparrow}} \chi_\uparrow \right.$$

$$\left. + ik_\downarrow t_\downarrow \sin\left(\frac{\theta}{2}\right) \frac{e^{ik_\downarrow x}}{\sqrt{k_\downarrow}} \chi_\downarrow \right)$$

$$= i \left[|t_\uparrow|^2 \cos^2\left(\frac{\theta}{2}\right) \chi^\dagger_\uparrow \sigma_1 \chi_\uparrow + \frac{\cos\left(\frac{\theta}{2}\right) \sin\left(\frac{\theta}{2}\right)}{\sqrt{k_\downarrow k_\uparrow}} \left(k_\downarrow t^*_\uparrow t_\downarrow e^{i(k_\downarrow - k_\uparrow)x} \chi^\dagger_\uparrow \sigma_1 \chi_\downarrow \right. \right.$$

$$\left. \left. + k_\uparrow t^*_\downarrow t_\uparrow e^{-i(k_\downarrow - k_\uparrow)x} \chi^\dagger_\downarrow \sigma_1 \chi_\uparrow \right) + |t_\downarrow|^2 \sin^2\left(\frac{\theta}{2}\right) \chi^\dagger_\downarrow \sigma_1 \chi_\downarrow \right]$$

$$= i \frac{\cos\left(\frac{\theta}{2}\right) \sin\left(\frac{\theta}{2}\right)}{\sqrt{k_\downarrow k_\uparrow}} \left(k_\downarrow t^*_\uparrow t_\downarrow e^{i(k_\downarrow - k_\uparrow)x} + k_\uparrow t^*_\downarrow t_\uparrow e^{-i(k_\downarrow - k_\uparrow)x} \right)$$

$$= i \frac{\sin\theta}{2\sqrt{k_\downarrow k_\uparrow}} \left(k_\downarrow t^*_\uparrow t_\downarrow e^{i(k_\downarrow - k_\uparrow)x} + k_\uparrow t^*_\downarrow t_\uparrow e^{-i(k_\downarrow - k_\uparrow)x} \right)$$

Note that

$$az + bz^* = a(x + iy) + b(x - iy)$$
$$= x(a + b) + iy(a - b)$$
$$= (a + b)\,\text{Re}(z) + i(a - b)\,\text{Im}(z).$$

Letting $z = t^*_\uparrow t_\downarrow e^{i(k_\downarrow - k_\uparrow)x}$, $a = k_\downarrow$, and $b = k_\uparrow$ gives

$$\tilde{J}^{\text{tr}}_{xx} = i \frac{\sin\theta}{2\sqrt{k_\downarrow k_\uparrow}} \left[(k_\downarrow + k_\uparrow)\,\text{Re}\left(t^*_\uparrow t_\downarrow e^{i(k_\downarrow - k_\uparrow)x} \right) + i(k_\downarrow - k_\uparrow)\,\text{Im}\left(t^*_\uparrow t_\downarrow e^{i(k_\downarrow - k_\uparrow)x} \right) \right].$$

Taking the imaginary part and using (10.24) gives

$$\boxed{J^{\text{tr}}_{xx} = \frac{\hbar^2}{2m} \frac{(k_\downarrow + k_\uparrow)}{2\sqrt{k_\downarrow k_\uparrow}} \sin\theta\,\text{Re}\left(t^*_\uparrow t_\downarrow e^{i(k_\downarrow - k_\uparrow)x} \right).}$$

Similarly

$$\tilde{J}^{\text{tr}}_{yx} = i \frac{\cos\left(\frac{\theta}{2}\right) \sin\left(\frac{\theta}{2}\right)}{\sqrt{k_\downarrow k_\uparrow}} \left(-ik_\downarrow t^*_\uparrow t_\downarrow e^{i(k_\downarrow - k_\uparrow)x} + ik_\uparrow t^*_\downarrow t_\uparrow e^{-i(k_\downarrow - k_\uparrow)x} \right)$$

$$= \frac{\sin\theta}{2\sqrt{k_\downarrow k_\uparrow}} \left(k_\downarrow t^*_\uparrow t_\downarrow e^{i(k_\downarrow - k_\uparrow)x} - k_\uparrow t^*_\downarrow t_\uparrow e^{-i(k_\downarrow - k_\uparrow)x} \right).$$

As before, let $z = t_\uparrow^* t_\downarrow e^{i(k_\downarrow - k_\uparrow)x}$, $a = k_\downarrow$, and $b = -k_\uparrow$:

$$\tilde{J}_{yx}^{tr} = \frac{\sin\theta}{2\sqrt{k_\downarrow k_\uparrow}} \left[(k_\downarrow - k_\uparrow) \operatorname{Re}\left(t_\uparrow^* t_\downarrow e^{i(k_\downarrow - k_\uparrow)x} \right) + i\left(k_\downarrow + k_\uparrow \right) \operatorname{Im}\left(t_\uparrow^* t_\downarrow e^{i(k_\downarrow - k_\uparrow)x} \right) \right].$$

Again taking the imaginary part and using (10.24) gives

$$J_{yx}^{tr} = \frac{\hbar^2}{2m} \frac{(k_\downarrow + k_\uparrow)}{2\sqrt{k_\downarrow k_\uparrow}} \sin\theta \operatorname{Im}\left(t_\uparrow^* t_\downarrow e^{i(k_\downarrow - k_\uparrow)x} \right).$$

Now to compute the last element, J_{zx}^{tr}

$$\tilde{J}_{zx}^{tr} = i\left[|t_\uparrow|^2 \cos^2\left(\frac{\theta}{2}\right) - |t_\downarrow|^2 \sin^2\left(\frac{\theta}{2}\right) \right],$$

and

$$J_{zx}^{tr} = \frac{\hbar^2}{2m} \left[|t_\uparrow|^2 \cos^2\left(\frac{\theta}{2}\right) - |t_\downarrow|^2 \sin^2\left(\frac{\theta}{2}\right) \right].$$

Putting it all together gives

$$\overline{J}^{tr} \cdot \hat{x} = \frac{\hbar^2}{2m} \left[\hat{x} \frac{(k_\downarrow + k_\uparrow)}{2\sqrt{k_\downarrow k_\uparrow}} \sin\theta \operatorname{Re}\left(t_\uparrow^* t_\downarrow e^{i(k_\downarrow - k_\uparrow)x} \right) \right.$$
$$+ \hat{y} \frac{(k_\downarrow + k_\uparrow)}{2\sqrt{k_\downarrow k_\uparrow}} \sin\theta \operatorname{Im}\left(t_\uparrow^* t_\downarrow e^{i(k_\downarrow - k_\uparrow)x} \right)$$
$$\left. + \hat{z}\left(|t_\uparrow|^2 \cos^2\left(\frac{\theta}{2}\right) - |t_\downarrow|^2 \sin^2\left(\frac{\theta}{2}\right) \right) \right].$$

Problem 10.3 Use the results of Problem 10.2 to verify (10.8)

$$\mathbf{N} = \frac{\hbar^2 \sin\theta}{2m} \left\{ \hat{x}\left[1 - \operatorname{Re}\left(r_\uparrow^* r_\downarrow + \frac{(k_\uparrow + k_\downarrow)}{2\sqrt{k_\uparrow k_\downarrow}} t_\uparrow^* t_\downarrow e^{i(k_\downarrow - k_\uparrow)x} \right) \right] \right.$$
$$\left. - \hat{y} \operatorname{Im}\left(r_\uparrow^* r_\downarrow + \frac{(k_\uparrow + k_\downarrow)}{2\sqrt{k_\uparrow k_\downarrow}} t_\uparrow^* t_\downarrow e^{i(k_\downarrow - k_\uparrow)x} \right) \right\} \qquad (10.26)$$

Solution 10.3
The areal torque density $\overline{N}$ is given by (10.7)

$$\overline{N} = \left[\overline{J}^{in} + \overline{J}^{ref} - \overline{J}^{tr} \right] \cdot \hat{x}.$$

Substituting the expressions from Problem 10.2 and collecting vector components gives

$$\frac{2m}{\hbar^2}\overline{N} = \hat{x}\sin\theta\left[1 - \mathrm{Re}\left(r_\uparrow^* r_\downarrow\right) - \frac{(k_\downarrow + k_\uparrow)}{2\sqrt{k_\downarrow k_\uparrow}}\mathrm{Re}\left(t_\uparrow^* t_\downarrow e^{i(k_\downarrow - k_\uparrow)x}\right)\right]$$

$$+ \hat{y}\sin\theta\left[-\mathrm{Im}\left(r_\uparrow^* r_\downarrow\right) - \frac{(k_\downarrow + k_\uparrow)}{2\sqrt{k_\downarrow k_\uparrow}}\mathrm{Im}\left(t_\uparrow^* t_\downarrow e^{i(k_\downarrow - k_\uparrow)x}\right)\right]$$

$$+ \hat{z}\left[\cos\theta - \cos^2\left(\frac{\theta}{2}\right)\left(|r_\uparrow|^2 + |t_\uparrow|^2\right) + \sin^2\left(\frac{\theta}{2}\right)\left(|r_\downarrow|^2 + |t_\downarrow|^2\right)\right].$$

But $|r_\uparrow|^2 + |t_\uparrow|^2 = |r_\downarrow|^2 + |t_\downarrow|^2 = 1$, and $\cos^2(\theta/2) - \sin^2(\theta/2) = \cos\theta$, so the z component becomes

$$\frac{2m}{\hbar^2}N_z = \cos\theta - \cos^2\left(\frac{\theta}{2}\right) + \sin^2\left(\frac{\theta}{2}\right) = \cos\theta - \cos\theta = 0.$$

The final result is therefore

$$\overline{N} = \frac{\hbar^2}{2m}\sin\theta\left\{\hat{x}\left[1 - \mathrm{Re}\left(r_\uparrow^* r_\downarrow + \frac{(k_\downarrow + k_\uparrow)}{2\sqrt{k_\downarrow k_\uparrow}}t_\uparrow^* t_\downarrow e^{i(k_\downarrow - k_\uparrow)x}\right)\right]\right.$$

$$\left. - \hat{y}\,\mathrm{Im}\left(r_\uparrow^* r_\downarrow + \frac{(k_\downarrow + k_\uparrow)}{2\sqrt{k_\downarrow k_\uparrow}}t_\uparrow^* t_\downarrow e^{i(k_\downarrow - k_\uparrow)x}\right)\right\},$$

in agreement with (10.8). Note that in writing this final step we have made use of the fact that $k_\uparrow$, $k_\downarrow$ are real.

10.2 Poynting Vector and Backward Waves

The complex fields of a uniform plane wave with *wave fronts* traveling in the $\hat{z}$ direction can be expressed as

$$\mathbf{e} = \hat{x}\,e_0 e^{i\mathbf{k}\cdot\mathbf{r}}, \quad \mathbf{h} = \hat{y}\,\frac{e_0}{\eta_0}e^{i\mathbf{k}\cdot\mathbf{r}}, \tag{10.27}$$

where $\mathbf{k} = \hat{z}k$ is the wave number and is pointing in the direction in which the phase fronts are moving. The velocity at which the constant phase fronts or planes move is referred to as the phase velocity.

As was defined in Sect. 4.3, the time-averaged power per unit area is given by the time-averaged Poynting vector

$$\mathbf{P} = \frac{1}{2}\mathrm{Re}\left(\mathbf{e}\times\mathbf{h}^*\right). \tag{10.28}$$

An application of the "right-hand rule" will quickly show that the power is also traveling along the $\hat{z}$ direction. We can, therefore, write

$$\mathbf{k} \cdot (\mathbf{e} \times \mathbf{h}^*) > 0. \tag{10.29}$$

When the power moves in the same direction as the constant phase planes, the wave is referred to as a *forward wave*.

From Maxwell's equation from Faraday's law, we see that this relationship between **e** and **h** depends on the sign and form of the permeability:

$$\mathbf{k} \times \mathbf{e} = \omega \overline{\mu} \cdot \mathbf{h}. \tag{10.30}$$

As an example, if the permeability is a negative scalar, then **e**, **h**, and **k** are related by the *"left-hand-rule."* Waves with the property

$$\mathbf{k} \cdot (\mathbf{e} \times \mathbf{h}^*) < 0 \tag{10.31}$$

are referred to as *backward waves* or *left-handed waves*.

In the case of a guided wave where the fields are not uniform across a constant phase plane, we integrate over a cross-section of the guided wave to find out if the mode is right-handed or left-handed (or forward or backward). Thus

$$\int_S \beta \cdot (\mathbf{e} \times \mathbf{h}^*) ds > 0 \quad \text{for forward waves;} \tag{10.32}$$

$$\int_S \beta \cdot (\mathbf{e} \times \mathbf{h}^*) ds < 0 \quad \text{for backward waves,} \tag{10.33}$$

where the integration is over the cross-section of the guided mode, and β is the wave vector of the mode.

Problem 10.4 The direction of the phase velocity is the same as the direction of the in-plane wave vector defined by $\beta = k_z \hat{z}$. To establish whether or not the wave is left-handed, we, therefore, examine the sign of the quantity $\beta \cdot (\mathbf{e} \times \mathbf{h}^*)$. If this quantity is negative, the transverse components of **e** and **h** necessarily form a left-handed triplet with β.

(a) By deriving the relevant lowest order backward volume wave expression for **e** and **h** for all y, show that

$$\beta \cdot \left(\mathbf{e} \times \mathbf{h}^*\right) < 0, \quad |y| \leq d/2 \tag{10.34}$$
$$\beta \cdot \left(\mathbf{e} \times \mathbf{h}^*\right) > 0, \quad |y| > d/2 \tag{10.35}$$

indicating left-handed fields within the film and right-handed fields outside the film.

(b) Show that

$$\int_{-\infty}^{\infty} \beta \cdot (\mathbf{e} \times \mathbf{h}^*)\, dy \; < \; 0, \tag{10.36}$$

and thus that the guided wave exhibits a net left-handed behavior.

Solution 10.4

The potential function for the lowest order (odd) backward volume spin wave mode
is given by

$$\psi(\bar{r}) = \begin{cases} \psi_0 \exp\left(\frac{k_z d}{2}\right) \sin\left(\frac{k_y d}{2}\right) \exp\left(i\nu k_z z - k_z y\right), & y > \frac{d}{2} \\ \psi_0 \sin\left(k_y y\right) e^{i\nu k_z z}, & |y| \leq \frac{d}{2} \\ -\psi_0 \exp\left(\frac{k_z d}{2}\right) \sin\left(\frac{k_y d}{2}\right) \exp\left(i\nu k_z z + k_z y\right), & y < -\frac{d}{2} \end{cases}$$

with $\nu = \pm 1$. The magnetic field intensity is obtained from $\bar{h} = -\nabla\psi$, from which
we obtain

$$\bar{h} = \begin{cases} k_z \psi\, \hat{y} - i\nu k_z \psi\, \hat{z}, & y > \frac{d}{2} \\ -k_z \psi_0 \cos\left(k_y y\right) e^{i\nu k_z z}\, \hat{y} - i\nu k_z z\psi\, \hat{z}, & |y| \leq \frac{d}{2} \\ k_z \psi\, \hat{y} + i\nu k_z \psi\, \hat{z}, & y < -\frac{d}{2} \end{cases}$$

Now given that

$$\bar{b} = \mu_0 \begin{bmatrix} (1+\chi) & -ik & 0 \\ ik & (1+\chi) & 0 \\ 0 & 0 & 1 \end{bmatrix} \begin{bmatrix} 0 \\ h_y \\ h_z \end{bmatrix},$$

we get

$$\begin{bmatrix} b_y \\ b_z \end{bmatrix} = \mu_0 \begin{bmatrix} (1+\chi)\, h_y \\ h_z \end{bmatrix}.$$

Consequently, the first-order electric field is given by

$$\nabla \times \bar{e} = i\omega\bar{b}$$

$$\partial_z e_x - \cancel{\partial_x e_z}^{0} = i\omega b_y \tag{10.37a}$$

$$\cancel{\partial_x e_y}^{0} - \partial_y e_x = i\omega b_z \tag{10.37b}$$

We will use (10.37a) to find e_x. It is left as an exercise to the reader to show that the
resulting expression also satisfies (10.37b).

(a) From (10.37a):

$$e_x = i\omega \int b_y dz = \frac{i\omega}{i\nu k_z} b_y = \frac{\omega}{\nu k_z} b_y.$$

For $y > \frac{d}{2}$,

$$e_x = \frac{\omega}{\nu k_z} \mu_0 h_y = \frac{\omega}{\nu k_z} \mu_0 k_z \psi = \omega \mu_0 \nu \psi.$$

For $|y| \leq \frac{d}{2}$

$$e_x = \frac{\omega}{\nu k_z} b_y = \frac{\omega}{\nu k_z} \mu_0 (1 + \chi) h_y = \frac{-\omega}{\nu k_z} (1 + \chi) k_y \psi_0 \cos(k_y y) e^{i\nu k_z z}$$

$$= \frac{-\omega \mu_0 k_y}{\nu k_z} (1 + \chi) \psi_0 \cos(k_y y) e^{i\nu k_z z}$$

For $y < \frac{-d}{2}$

$$e_x = \frac{\omega}{\nu k_z} b_y = \frac{\omega}{\nu k_z} \mu_0 h_y = \frac{\omega}{\nu k_z} \mu_0 k_z \psi = \omega \mu_0 \nu \psi.$$

To summarize

$$e_x = \begin{cases} \omega \mu_0 \nu \psi \,, & |y| > \frac{d}{2} \\[2mm] \dfrac{-\omega \mu_0 k_y}{\nu k_z} (1 + \chi) \psi_0 \cos(k_y y) e^{i\nu k_z z} \,, & |y| \leq \frac{d}{2} \end{cases}$$

We are now ready to evaluate $\bar{\beta} \cdot (\bar{e} \times \bar{h}^*)$, where $\bar{\beta} = \hat{z} \nu k_z$. This gives

$$\nu k_z \hat{z} \cdot (\bar{e} \times \bar{h}^*) = \nu k_z \left(e_x h_y^* - e_y h_x^* \right)$$

$$= \nu k_z e_x h_y^*$$

The last step follows since the first-order electric field only has an $\hat{x}$ component. For $|y| \leq \frac{d}{2}$

$$\bar{\beta} \cdot (\bar{e} \times \bar{h}^*) = \nu k_z e_x h_y^*$$

$$= \nu k_z \left(\frac{-\omega \mu_0 k_y}{\nu k_z} (1 + \chi) \psi_0 \cos(k_y y) e^{i\nu k_z z} \right)$$

$$\times \left(-k_y \psi_0 \cos(k_y y) e^{-i\nu k_z z} \right)$$

$$= \omega \mu_0 k_y^2 (1 + \chi) \psi_0^2 \cos^2(k_y y).$$

Walker's equation for a plane wave excitation is given by (see (5.15) and in [6, Chap. 5])

$$(1 + \chi)(k_x^2 + k_y^2) + k_z^2 = 0. \tag{10.38}$$

In the present case $k_x = 0$, so

$$k_y^2 (1 + \chi) = -k_z^2.$$

Therefore

$$\boxed{\bar{\beta} \cdot (\bar{e} \times \bar{h}^*) = -\omega \mu_0 k_z^2 \psi_0^2 \cos^2 (k_y y) < 0 \quad \text{for} \quad |y| \leq d/2}$$

For $|y| > d/2$

$$\bar{\beta} \cdot (\bar{e} \times \bar{h}^*) = \nu k_z e_x h_y^*$$
$$= \nu k_z \omega \mu_0 \nu \psi k_z \psi^*$$
$$= \nu^2 \omega \mu_0 k_z^2 |\psi|^2$$
$$= \omega \mu_0 k_z^2 |\psi|^2$$

In the last step, we used the fact that $\nu^2 = 1$. We, therefore, conclude that

$$\boxed{\bar{\beta} \cdot (\bar{e} \times \bar{h}^*) = \omega \mu_0 k_z^2 |\psi|^2 > 0 \quad \text{for} \quad |y| > d/2}$$

(b) We now integrate the quantities from Part (a) over all y

$$\int_{-\infty}^{\infty} \bar{\beta} \cdot (\bar{e} \times \bar{h}^*) \, dy$$

$$= \omega \mu_0 k_z^2 \psi_0^2 \left[\int_{-\infty}^{-d/2} \frac{|\psi|^2}{\psi_0^2} dy - \int_{-d/2}^{d/2} \cos^2 (k_y y) \, dy + \int_{d/2}^{\infty} \frac{|\psi|^2}{\psi_0^2} dy \right]$$

that we rewrite as

$$\frac{1}{\omega \mu_0 k_z^2 \psi_0^2} \int_{-\infty}^{\infty} \bar{\beta} \cdot (\bar{e} \times \bar{h}^*) \, dy = I_1 - I_2 + I_3.$$

First consider I_1

$$I_1 = \int_{-\infty}^{-d/2} \frac{|\psi|^2}{\psi_0^2} dy = \exp(k_z d) \sin^2 \left(k_y \frac{d}{2} \right) \frac{\exp(-k_z d)}{2 k_z} = \frac{\sin^2 (k_y d/2)}{2 k_z}.$$

Similarly

$$I_3 = \exp(k_z d) \sin^2\left(k_y \frac{d}{2}\right) \int_{d/2}^{\infty} \exp(-2k_z y)\, dy.$$

Thus

$$I_3 = I_1 = \frac{1}{2k_z} \sin^2\left(\frac{k_y d}{2}\right).$$

Finally

$$I_2 = \int_{-d/2}^{d/2} \cos^2(k_y y)\, dy = \frac{1}{k_y} \int_{-k_y d/2}^{k_y d/2} \cos^2(\theta)\, d\theta$$

$$= \frac{1}{2k_y} \left[\theta + \sin\theta\cos\theta\right]_{-k_y d/2}^{k_y d/2}$$

$$= \frac{1}{2k_y} \left[k_y d + 2\sin\left(k_y \frac{d}{2}\right)\cos\left(k_y \frac{d}{2}\right)\right]$$

Putting it all together

$$\frac{1}{\omega\mu_0 k_z^2 \psi_0^2} \int_{-\infty}^{\infty} \bar{\beta}\cdot(\bar{e}\times\bar{h}^*)\, dy$$

$$= \frac{\sin^2\left(k_y \frac{d}{2}\right)}{k_z} - \frac{1}{2k_y}\left[k_y d + 2\sin\left(k_y \frac{d}{2}\right)\cos\left(k_y \frac{d}{2}\right)\right] \qquad (10.39)$$

$$= \frac{-d}{2} + \frac{\sin\left(k_y \frac{d}{2}\right)\cos\left(k_y \frac{d}{2}\right)}{k_z}\left[\tan\left(k_y \frac{d}{2}\right) - \frac{k_z}{k_y}\right]$$

From Walker's equation (10.38), we have

$$\frac{k_z}{k_y} = \sqrt{-(1+\chi)}.$$

Substituting this into the bracketed equation on the right of (10.39) gives

$$\frac{1}{\omega\mu_0 k_z^2 \psi_0^2} \int_{-\infty}^{\infty} \bar{\beta}\cdot(\bar{e}\times\bar{h}^*)\, dy$$

$$= \frac{-d}{2} + \frac{\sin\left(k_y \frac{d}{2}\right)\cos\left(k_y \frac{d}{2}\right)}{k_z} \times \left[\tan\left(\frac{k_z d}{2\sqrt{-(1+\chi)}}\right) - \sqrt{-(1+\chi)}\right]$$
$$(10.40)$$

But from the backward volume wave dispersion relation for the lowest order mode (5.46a), and [6, Chap. 5], we know that

$$\tan\left[\frac{k_z d}{2\sqrt{-(1+\chi)}}\right] = \sqrt{-(1+\chi)}.$$

Thus, the expression in the bracket on the right-hand side of (10.40) vanishes, and we have

$$\int_{-\infty}^{\infty} \bar{\beta} \cdot \left(\bar{e} \times \bar{h}^*\right) dy = -\omega\mu_0 k_z^2 \psi_0^2 \frac{d}{2} < 0.$$

Therefore, we conclude that the mode has net left-handed behavior.

References

1. R.K. Dumas, A. Houshang, J. Akerman, Propagating spin waves in nanocontact spin torque oscillators, in *Spin Wave Confinement: Propagating Waves*, eds. by S.E. Demokritov (Pan Stanford Publishing, Singapore, 2017)
2. J.C. Slonczewski, Current-driven excitation of magnetic multilayers. J. Magn. Magn. Matl. **159**, L1 (1996)
3. L. Berger, Emission of spin waves by a magnetic multilayer traversed by a current. Phys. Rev. B **54**, 9353 (1996)
4. M. Stiles, A. Zangwill, Anatomy of spin-transfer torque. Phys. Rev. B **66**, 014407 (2002)
5. E. Merzbacher, *Quantum Mechanics*, 3rd edn. (Wiley, New York, 1998)
6. D.D. Stancil, A. Prabhakar, *Spin Waves: Theory and Applications* (Springer, New York, 2009)

Appendix

© The Editor(s) (if applicable) and The Author(s), under exclusive license to Springer   241
Nature Switzerland AG 2021
D. D. Stancil and A. Prabhakar, *Spin Waves*,
https://doi.org/10.1007/978-3-030-68582-9

Table 1 Organization of problems

Topic	Spin Waves	Solutions
Spinning Top	1.1	1.1
Angular Momentum Operator	1.2 - 1.5	1.2 - 1.5
Hund's Rules	1.6 - 1.8	1.6 - 1.8
Coupled Spins	2.1	2.1
Harmonic Oscillator	2.2	2.2
Spin Raising and Lowering Operators	-	2.3
"	2.3	2.4
Spin Waves	2.4 - 2.5	2.5 - 2.6
Magnetic Materials	3.1 - 3.3	3.1 - 3.3
Miller Indices, Magnetocrystalline Anisotropy	3.4 - 3.7	3.4 - 3.7
Magnetic Susceptibility	3.8 - 3.10	3.8 - 3.10
Stoner-Wolfarth Model	3. 11	3. 11
Plane Waves	4.1 - 4.3	4.1 - 4.3
"	-	4.4
"	4.4 - 4.5	4.5 - 4.6
Magnetostatic Approximation	4.6	5.1
Uniform Precession Modes	5.2 - 5.7	5.2 - 5.7
Exchange-dominated Spin Waves	5.1	5.8
Spin Relaxation	6.1 - 6.3	6.1 - 6.3
Mode Orthogonality & Normalization	6.4 - 6.6	6.4 - 6.6
Radiation Resistance	6.7 - 6.9	6.7 - 6.9
Calculus of Variations	7.x	7.x
Optical Guided Modes	8.1 - 8.6	8.1 - 8.6
Magnetic and Electric Susceptibility	8.7 - 8.10	8.7 - 8.10
Coupled Mode Theory, Magneto-optics	8.11 - 8.15	8.11 - 8.15
Complex Spin Wave Amplitudes	9.1 - 9.3	9.1 - 9.3
Bogoliubov Transformation	9.4 - 9.5	9.4 - 9.5
Nonlinear Schrödinger Equation	9.6 - 9.8	9.6 - 9.8
Spin Transfer Torque	10.1 - 10.3	10.1 - 10.3
Poynting Vector for Backward Waves	10.4	10.4

Index

© The Editor(s) (if applicable) and The Author(s), under exclusive license to Springer
Nature Switzerland AG 2021
D. D. Stancil and A. Prabhakar, *Spin Waves*,
https://doi.org/10.1007/978-3-030-68582-9

Printed in the United States
by Baker & Taylor Publisher Services